Wind and Wave Energy Resource Assessment and Combined Utilization

Wind and Wave Energy Resource Assessment and Combined Utilization

Editors

Guanghua He
Liang Sun
Yan Bao

Basel • Beijing • Wuhan • Barcelona • Belgrade • Novi Sad • Cluj • Manchester

Editors
Guanghua He
Harbin Institute of Technology
Weihai
China

Liang Sun
Wuhan University of Technology
Wuhan
China

Yan Bao
Shanghai Jiao Tong University
Shanghai
China

Editorial Office
MDPI AG
Grosspeteranlage 5
4052 Basel, Switzerland

This is a reprint of articles from the Special Issue published online in the open access journal *Energies* (ISSN 1996-1073) (available at: https://www.mdpi.com/journal/energies/special_issues/Wind_Wave_Assessment_Utilization).

For citation purposes, cite each article independently as indicated on the article page online and as indicated below:

Lastname, A.A.; Lastname, B.B. Article Title. *Journal Name* **Year**, *Volume Number*, Page Range.

ISBN 978-3-7258-2581-3 (Hbk)
ISBN 978-3-7258-2582-0 (PDF)
doi.org/10.3390/books978-3-7258-2582-0

© 2024 by the authors. Articles in this book are Open Access and distributed under the Creative Commons Attribution (CC BY) license. The book as a whole is distributed by MDPI under the terms and conditions of the Creative Commons Attribution-NonCommercial-NoDerivs (CC BY-NC-ND) license.

Contents

Luis Baquero, Herena Torio and Paul Leask
Machine Learning Algorithms for Vertical Wind Speed Data Extrapolation: Comparison and Performance Using Mesoscale and Measured Site Data
Reprinted from: *Energies* 2022, 15, 5518, doi:10.3390/en15155518 1

Ruijia Jin, Jiawei Wang, Hanbao Chen, Baolei Geng and Zhen Liu
Numerical Investigation of Multi-Floater Truss-Type Wave Energy Convertor Platform
Reprinted from: *Energies* 2022, 15, 5675, doi:10.3390/en15155675 21

Xianxiong Zhang, Bin Li, Zhenwei Hu, Jiang Deng, Panpan Xiao and Mingsheng Chen
Research on Size Optimization of Wave Energy Converters Based on a Floating Wind-Wave Combined Power Generation Platform
Reprinted from: *Energies* 2022, 15, 8681, doi:10.3390/en15228681 38

Zhongxian Chen, Xu Li, Yingjie Cui and Liwei Hong
Modeling, Experimental Analysis, and Optimized Control of an Ocean Wave Energy Conversion System in the Yellow Sea near Lianyungang Port
Reprinted from: *Energies* 2022, 15, 8788, doi:10.3390/en15238788 54

Hongyuan Sun, Jiazheng Wang, Haihua Lin, Guanghua He, Zhigang Zhang, Bo Gao and Bo Jiao
Numerical Study on a Cylinder Vibrator in the Hydrodynamics of a Wind–Wave Combined Power Generation System under Different Mass Ratios
Reprinted from: *Energies* 2022, 15, 9265, doi:10.3390/en15249265 70

Mahsa Dehghan Manshadi, Milad Mousavi, M. Soltani, Amir Mosavi and Levente Kovacs
Deep Learning for Modeling an Offshore Hybrid Wind–Wave Energy System
Reprinted from: *Energies* 2022, 15, 9484, doi:10.3390/en15249484 86

Shabnam Hosseinzadeh, Amir Etemad-Shahidi and Rodney A. Stewart
Site Selection of Combined Offshore Wind and Wave Energy Farms: A Systematic Review
Reprinted from: *Energies* 2023, 16, 2074, doi:10.3390/en16042074 102

Amar Azhar and Huzaifa Hashim
A Review of Wind Clustering Methods Based on the Wind Speed and Trend in Malaysia
Reprinted from: *Energies* 2023, 16, 3388, doi:10.3390/en16083388 135

Yingjie Cui, Fei Zhang and Zhongxian Chen
Predication of Ocean Wave Height for Ocean Wave Energy Conversion System
Reprinted from: *Energies* 2023, 16, 3841, doi:10.3390/en16093841 159

Yongxu Jiang, Chenze Cao, Ting Cui, Hao Yang and Zhengjun Tian
Numerical Study on Auxiliary Propulsion Performance of Foldable Three-Element Wingsail Utilizing Wind Energy
Reprinted from: *Energies* 2024, 17, 3833, doi:10.3390/en17153833 172

Article

Machine Learning Algorithms for Vertical Wind Speed Data Extrapolation: Comparison and Performance Using Mesoscale and Measured Site Data

Luis Baquero [1,*], Herena Torio [1] and Paul Leask [2]

[1] Institute of Physics, Faculty of Mathematics and Science, Carl von Ossietzky University of Oldenburg, 26129 Oldenburg, Germany; herena.torio@uni-oldenburg.de
[2] Project Development and Analytics, Energy Systems, DNV, 26129 Oldenburg, Germany; paul.leask@dnv.com
* Correspondence: luis.baquero1@uni-oldenburg.de

Abstract: Machine learning (ML) could be used to overcome one of the largest sources of uncertainty in wind resource assessment: to accurately predict the wind speed (WS) at the wind turbine hub height. Therefore, this research defined and evaluated the performance of seven ML supervised algorithms (regressions, decision tree, support vector machines, and an ensemble method) trained with meteorological mast data (temperature, humidity, wind direction, and wind speeds at 50 and 75 m), and mesoscale data below 80 m (from the New European Wind Atlas) to predict the WS at the height of 102 m. The results were compared with the conventional method used in wind energy assessments to vertically extrapolate the WS, the power law. It was proved that the ML models overcome the conventional method in terms of the prediction errors and the coefficient of determination. The main advantage of ML over the power-law was that ML performed the task using either only mesoscale data (described in scenario A), only data from the measurement mast (described in scenario B) or combining these two data sets (described in scenario C). The best ML models were the ensemble method in scenario A with an R^2 of 0.63, the linear regression in scenario B with an R^2 of 0.97, and the Ridge regressor in scenario C with an R^2 of 0.97.

Keywords: wind speed extrapolation; power-law; machine learning; supervised learning; mesoscale model; wind energy; energy production assessment; new european wind atlas; random forest; support vector machines; linear regression

1. Introduction

1.1. Motivation and Incitement

The growing problem of climate change, energy security, access to energy, and unstable oil and gas prices, accelerated the energy transition to low-carbon technology options such as renewable energies. According to the IRENA [1], the world cumulative installed capacity of renewable energies rose to 2536 GW by the end of 2019. Wind energy occupied the second place, after hydropower, in the largest installed capacity green technology, with 24.54% (calculated from [1]) and is expected to play a role of highest relevance in the transition of our energy systems towards renewable energy sources [2]. To pave the way for such a major increase in the installed wind capacity, reliable and low-cost wind resource site assessments are a must since they are the basis for a solid analysis of a particular location's economic and technical wind-exploitation potential.

The uncertainty present in all wind resource estimates is currently most commonly related to the following factors: wind speed measurements, the historical climate adjustment, potential future climate deviations, vertical extrapolation of wind speed measurements to the proposed hub height, and the spatial wind resource distribution [3]. Gasch [4] estimates that an error of 10% in the wind speed measurements may produce an error in the determined power output of up to 33%. Furthermore, calculating the wind regime on a

site as exact as possible is also beneficial for determining the mechanical loads and stress. To obtain the best possible description of the wind resource, a measurement campaign is performed using meteorological masts and/or ground-based remote sensing systems. Since meteorological towers are often shorter than a turbine's hub height, it is necessary to extrapolate speed measurements to higher heights. This task requires a careful and often subjective analysis of the mast and site information, including the observed shear, local meteorology, topography, and land cover.

Machine learning algorithms—tough complex and computational intensive—are powerful and accurate tools for extrapolating wind speed data [5,6]. The primary benefit of using machine learning algorithms to extrapolate the wind speed to higher heights is the expected higher accuracy in the the resulting wind speed time series, due to the prediction model learning not only from representative wind speed data sets (recorded and modeled), but also from modeled information at the target height related to the general climate conditions such as humidity, temperature, wind direction, friction velocity, inverse Obukhov length, planetary boundary layer height, the surface latent heat flux, and solar radiation. Additionally, acquisition and use of information about the climate conditions from existing sources (e.g., New European Wind Atlas (NEWA) [7]) for a specific site within Europe has significantly lower costs than additional measurement sensors on site. In this paper, we create seven supervised learning models to obtain a wind speed time series at the height of 102 m, using recorded and mesoscale data from lower heights (75 and 50 m). Then, we assess the performance of the models and compare results with the power-law procedure [8] to compare whether the machine learning method outperforms the traditional approach. The models are based on the following algorithms: Linear regression, ridge regression, lasso regression, elastic regression, support vector machines, decision tree, and random forest.

1.2. Literature Review

In recent years, there has been considerable growing interest in applying machine learning in the field of wind energy, focused primarily on production forecasting, such as in [9–12]. Predicting power generation involves mainly wind speed forecasting, an increasing number of studies have found that using ML, wind assessments tasks can be conducted such as short-term and long-term wind speed predictions [13–17]. Few studies have been focused on the wind speed vertical extrapolation in the temporal domain. Cheggaga [18] studied the possibility of the use of an artificial neural network (ANN) to predict the wind speed at 50 m. He draws our attention to the use of the temperature time series to improve the model performance. The best performance was a Root mean square error (RMSE) of 5.171 m s^{-1} over a year data. Turkan [19] trained seven machine learning methods to predict the WS at 30 m above the ground and compared the results between them. Support vector machine (SVM) was the one who showed the best performance among the others. Cheggaga [20] improved his previous ANN model [18] by adding information from the power-law exponent, along with the temperature and wind speeds. The vertical extrapolation task was from 10, 30 m to 50 m high, and the modified ANN model obtained an RMSE of 0.87 m s^{-1}, which was around 50% better than Power-Law's. It confirmed our idea of incorporating into the training sets mathematical transformations based on the traditional methods such as the power law. Valsaraj [5] used an SVM to extrapolate the WS at 80 m and compare it against measured values. This model returned a RMSE below 1.5 m s^{-1} for all the predicted data points. Mohandes [6] applied a Deep neural network (DNN), an ANN, and a physical method to extrapolate LiDAR wind speed data to 120 m heights. In the analysis, a connection between the heights of measurements and the accuracy of the model was found. According to the Mean absolute percent error (MAPE), the model with the best performance was the DNN, followed by the traditional method, and the ANN in the third place. In a major advance in 2020, studies have investigated ML solutions in the temporal domain and the spatial domain and the use of non-dimensional features. Vassallo [21] used LiDAR data and an ANN to assess WS extrapolation in three different

terrain conditions. He found that using non-dimensional features improved the network performance up to 65% comparing with a traditional method. Bodini [22] trained a Random Forest model in vertical wind speed extrapolation task with a site LiDAR data and then use the model to predict in another four different locations. The random forest reduced the Mean absolute error MAE up 35% comparing with the traditional methods when the model is tested in the same location where it was trained and by 20% in the other sites. The degradation of the model was due to the modification of the geographical location. However, it still outperforms traditional methods. Adli [23] who evaluated a model that includes Response surface methodology (a set of statistical and mathematical techniques useful for the development, improvement, and optimization of processes) to extrapolate the wind speed from 10 m height up to 50 m height. He used WS and temperature data from a meteorological mast and could match a relation between these two variables to improve the predictions done by the ANN model in [20]. The experiment results show a better model performance ($R^2 = 0.99$) than the ANN model and the power-law method. This finding supports our research in exploring different ML models rather than the ANN and training the models not only with wind speed data, but also with the atmospheric variables from the mesoscale model. The last related work that dates from this year is by Emeksiz [24], where he explored a tree-based genetic programming algorithm to solve a vertical wind extrapolation task from 50 m to 100 m. The model outperformed the power-law, and the logarithmic-law (another method to vertically extrapolate wind speeds), by a reduction of 58% in their RMSE and outperformed an ANN model moving from 0.123 to 0.079 $m\,s^{-1}$ in the RMSE for the target site. The random forest and the tree-based genetic programming algorithm start from similar concepts in the way the decisions are generated. Emeksiz's work findings support that we have chosen an ensemble method based on the Random Forest since it is estimated that these algorithms have advantages in solving tasks in nonlinear and complex systems, and can even be combined to generate more robust models [23,25,26].

1.3. Major Contributions and Organization

This work presents two main contributions, the first is the comparison of different algorithms in solving the same task, and the second is the use of modeled data from mesoscale atmospheric models. The use of ANN opened the door to think of new methods to improve the predictions made with the power law. However, it has recently been shown that other algorithms such as deep learning or genetic algorithm approaches can also optimize the predictions of vertical wind speed extrapolation. In this aspect, it is of great importance to investigate and compare the most representative supervised learning algorithms in the solution of the same task, which would allow comparisons of their different characteristics, such as their computational cost, and their applicability to the target task. Additionally, the literature review shows that when algorithms are provided with additional information besides wind speed, predictions are improved. On this point, this work is novel in that it proposes the use of modeled data as part of the training set, which allows the models to take into account characteristics of all atmospheric conditions of the study site, in addition to opening the possibility of performing a feature engineering process that had a direct impact on the performance of the models and their results.

Section 2 of this report presents the selection of algorithms evaluated, the methodology used to build the models and experiments, the training data set, three scenarios for testing the algorithms, and how their results were evaluated. Section 3 presents the results of the experiments, and in Section 4, we discuss their implications.

2. Theory and Methods
2.1. Considered ML Approaches

Machine learning approaches have been widely investigated for solving tasks in the wind energy sector, especially supervised learning, because the target variable to predict or classify depends strictly on the nature of a predefined input data [27]. The

objective task of this research was also subjected to a supervised regression solution since here continuous numerical values are predicted from predefined data sets. These data samples are prepared and analyzed beforehand, the features are well-defined, and the labels/responses are already known in advance. Therefore, a supervised learning algorithm can construct relations and associations between the inputs and their corresponding outputs in the learning phase. Then, we can use those trained models to predict an output from unseen data.

The general procedure in a supervised learning task is presented in Figure 1. The methodology that rules our research is based on the Cross Industrial Standard Process for Data Mining, CRISP-DM [28] since it estimates, creates, evaluates, and redefines over and over, the machine learning system until getting a satisfactory result of the proposed models.

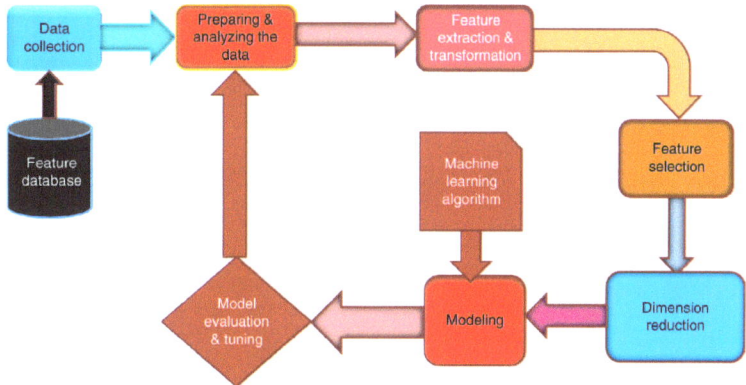

Figure 1. Supervised machine learning framework [29].

The models were trained using the same input data, taking special care in avoiding data-leakage in the train/split process. This allowed some observations of the advantages and disadvantages of each model for the first interactions. For subsequent model iteration, adjustments were made to the feature selections and their transformations to achieve the best results for each model. With this, we were able to identify the strengths of each model according to the training information and features used. The following algorithms were used to construct the models:

- Regression: Linear, Ridge, Lasso, and Elastic.
- Decision-making: Decision Trees (DT).
- Support vector machines: Support Vector Regression (SVR).
- Ensemble methods: Random Forest (RF).

A detailed theory related to these algorithms is available in [30].

These models were tested in three scenarios (refer to Section 2.3) in order to assess the performance, where they were subjected to different training sets. In the modeling process, several procedures were performed to reduce the uncertainties, such as data coverage, data quality, data imputation, time series analysis, statistical analysis, outliers analysis, feature engineering, hyperparameters tuning, cross-validation, and scoring evaluation. This process was carried out using the following libraries: Apache Spark [31], Scikit-learn [32], statsmodels [33], and pandas [34].

2.2. Training Data Sources

Two sources of data were available to train and assess the models. The first comes from a meteorological mast, henceforth named observed data, and the second one is modeled mesoscale data from the New European Wind Atlas (NEWA) [7], hereafter mesoscale data.

The observed data provided the following information: wind speed from a cup anemometer at 102.2 m, 75 m and 50 m; wind direction at 98.7 m, and 48.7. Temperature at 100 m, and 25 m; Pressure at 96.4 m; Relative humidity at 96.4 m. This data set has a resolution of 10 min average for each variable, and the time frame was two years. The signals at 50 m and 75 m were used to calculate the resulting wind speed time series at a height of 102.2 m. The results of the ML models and the power-law method, were compared against the measured wind speed at 102.2 m.

The NEWA mesoscale data, for the same location as the observed data, provides time series for 23 climate signals [35]. However, some of them are not related to the horizontal wind speed because they are not physically related to the vertical fluxes of heat, moisture, momentum, or roughness of the Earth's surface in any case [36]. Consequently, the best procedure was to discard those variables to avoid confusion in the models by giving information not related to the target feature. The variables listed in Table 1 were chosen and later studied to conclude whether those variables increase the model performance or not. The machine learning model for predicting wind speed is weather-dependent with seasonal variations. Therefore, the time frame of the data set should be sufficient to cover the seasonal variations: summer, winter, autumn, and spring.

Table 1. Selected features from the mesoscale data.

Item	Variable Name	Units	Nomenclature
1	Wind speed	$m\,s^{-1}$	WS
2	Wind Direcction	°	WD
3	Air Temperature	°C	T
4	Friction velocity	$m\,s^{-1}$	UST
5	Shortwave direct normal radiation	$W\,m^{-2}$	SWDDNI
6	Shortwave diffuse incident radiation	$W\,m^{-2}$	SWDDRI
7	Inverse Obukhov length	m^{-1}	RMOL
8	Planetary boundary layer height	m	PBLH
9	Surface pressure	Pa	PSFC
10	Surface latent Heat Flux	$W\,m^{-2}$	LH
11	Water vapour mixing ratio	$kg\,kg^{-1}$	QVAPOR
12	Turbulent kinetic energy	$m^2\,s^2$	TKE

2.3. Scenarios

Defining several scenarios brings the possibility to compare and analyze the cost/benefit of each model solution in terms of maintainability, computational demand, and accuracy [37]. The following scenarios allowed us to assess the performance of the seven models against the mesoscale data from the NEWA, the observed data from the met mast, and a combination of both.

- Scenario A: only mesoscale data is available to extrapolate the wind speed at 102 m height.
- Scenario B: only observed data is available to extrapolate the wind speed at 102 m height.
- Scenario C: observed, and mesoscale data are available to extrapolated wind speed at 102 m height.

2.4. Assessment

The learning model performance evaluation is used to assess the target approximation's quality that the model represents. It measures how close/far off the predicted values are versus the real values recorded by the met mast at a 102 m height. The following indicators were used to compare model performance (they were computed using 5 folds cross-validation). Explanations and significance for each metric are presented in Appendix B.

- Mean Absolute Error (MAE);
- Mean Squared Error (MSE);

- Root Mean Squared Error (RMSE);
- Coefficient of determination (R^2).

3. Results

3.1. Scenario A: Extrapolate Wind Speed Using Mesoscale Data from Newa

Statistical variables for assessing the models in this scenario are listed in Table 2, and their configurations in Figure A1. A graphical representation of the model's errors and the model's performance are presented in Figures 2 and 3, respectively.

Table 2. Results of the models when they are trained with NEWA mesoscale data.

Model	MAE	MSE	RMSE	R^2 CV
Linear	1.288161	2.907878	1.70525	0.626386
Ridge	1.274707	2.86232	1.691839	0.627854
Lasso	1.273021	2.853636	1.689271	0.62787
ElasticNet	1.271784	2.846432	1.687137	0.627652
Desicion Tree	1.565074	4.166773	2.041268	0.408519
SVR	1.288161	2.907878	1.70525	0.626386
Random Forest	1.269825	2.819217	1.679052	0.633857

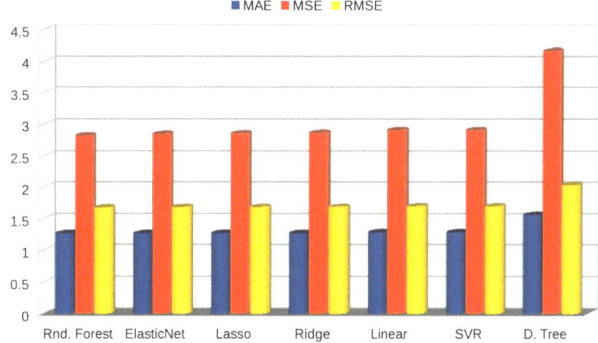

Figure 2. Model errors [m s^{-1}].

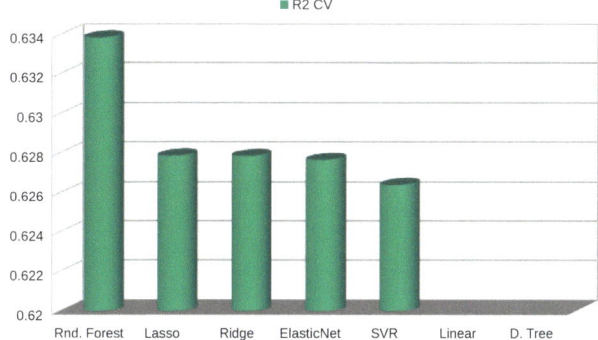

Figure 3. Performance scores.

The best wind speed predictions, at the target height and using only mesoscale data, were obtained by the model based on a Random Forest. It substantially exceeds the predictions obtained by the other ML models, and the power law. The scores obtained by

the RF were: MAE = 1.269, MSE = 2.819, RMSE = 1.679, and R^2 = 0.633. Additionally, the RF model was trained with the same information that the power law used, WS at 50 m, and 75 m height. It confirmed that the ML model is more robust than the power law, even when both models use the same available information. The metrics for the Power Law method are MAE = 1.555, MSE = 4.118, RMSE = 2.029, and R^2 = 0.446. Figure 4 shows the correlation of WS predicted by the RF with the data recorded by the met mast; Figure 5 shows an example of the RF predicted time series in a period of high variability of the wind speed.

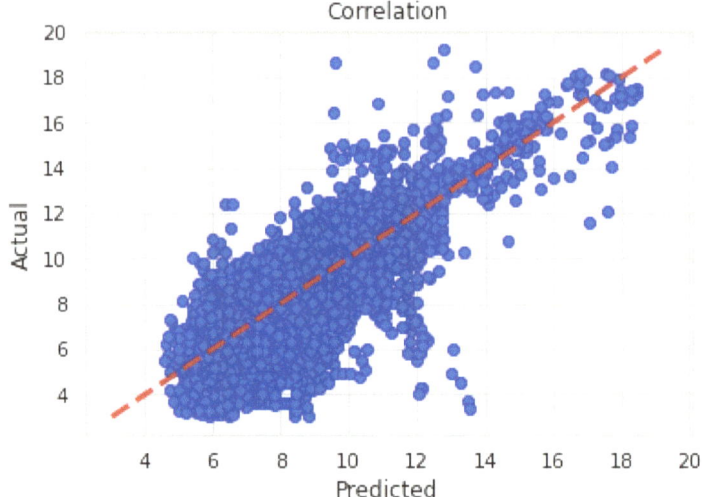

Figure 4. Correlation between observed (measured) data and results from the RF trained only with mesoscale data in m s^{-1}.

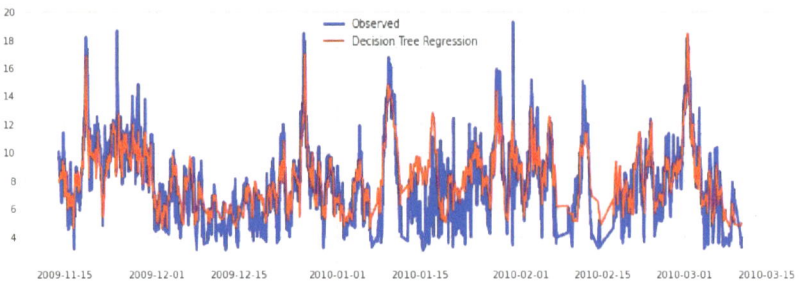

Figure 5. Observed and predicted wind speeds using RF based only in mesoscale data in m s^{-1}.

3.2. Scenario B: Extrapolate Wind Speeds Using Data from a Met Mast

Statistical variables for assessing the models in this scenario are listed in Table 3, and their configurations in Figure A2. A graphical representation of the model's errors and the model's performance are presented in Figures 6 and 7, respectively.

Table 3. Results of the models when they are trained with measured data.

Model	MAE	MSE	RMSE	R^2 CV
Linear	0.259943	0.283292	0.532251	0.97642
Ridge	0.259976	0.283458	0.532408	0.975386
Lasso	0.25996	0.283632	0.532571	0.975386
ElasticNet	0.259969	0.283562	0.532505	0.975386
D. Tree	0.279097	0.304115	0.551466	0.970416
SVR	0.24582	0.284564	0.533445	0.977656
Rnd. Forest	0.236336	0.278178	0.527425	0.97876

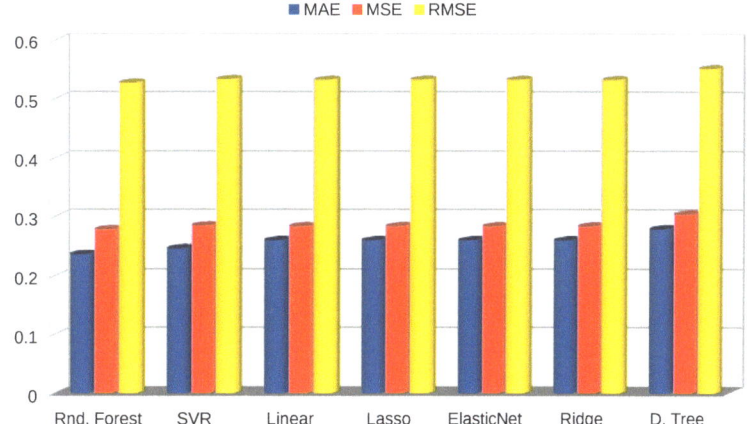

Figure 6. Model errors [m s^{-1}].

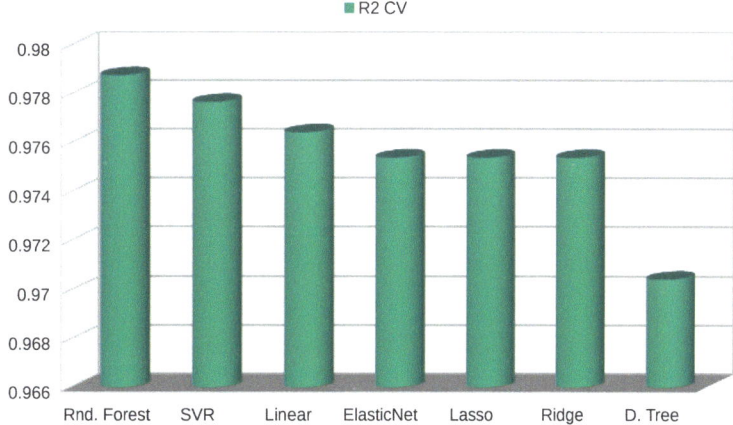

Figure 7. Performance scores.

Similarly to Scenario A, the best model performances in terms of coefficient of correlation and errors, using only observed data during the training phase, was the Random Forest (RF), followed by the Support vector machine, and then by the Linear regression. The RF obtained a MAE = 0.23633, MSE = 0.27817, RMSE = 0.52742, R^2 = 0.97876. All machine learning models outperformed the power law method, which metrics are MAE = 0.25359, MSE = 0.30319, RMSE = 0.55063, R^2 = 0.95922. Figure 8 shows the

correlation of WS predicted by the RF with the recorded by the met mast; Figure 9 shows an example of the RF predicted time series in a period of high variability of the wind speed.

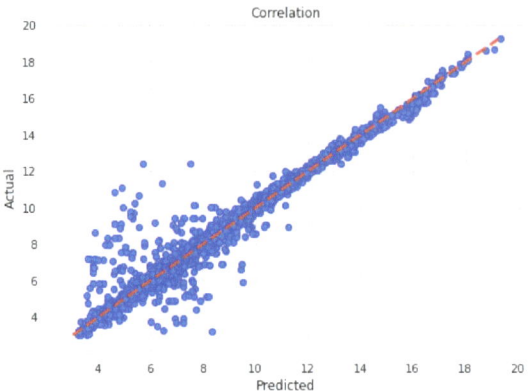

Figure 8. Correlation between observed (measured) data and results from the RF trained only with observed data in m s^{-1}.

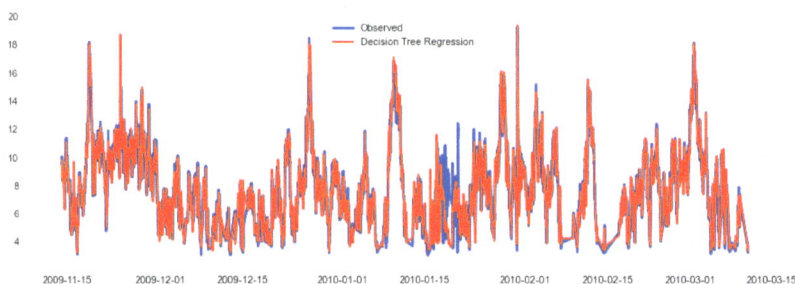

Figure 9. Observed and predicted wind speeds using the RF based only in mesoscale data in m s^{-1}.

3.3. Scenario C: Extrapolate Wind Speeds Using Data from a Met Mast and Mesoscale from the Newa

Statistical variables for assessing the models in this scenario are listed in Table 4, and their configurations in Figure A3. A graphical representation of the model's errors and the model's performance are presented in Figures 10 and 11, respectively.

Table 4. Results of the models when they are trained with mesoscale and observed data.

Model	MAE	MSE	RMSE	R^2 CV
Linear	0.256397	0.281716	0.530769	0.976455
Ridge	0.255516	0.280674	0.529786	0.976485
Lasso	0.256068	0.281552	0.530615	0.976416
ElasticNet	0.259969	0.283562	0.532505	0.975386
D. Tree	0.317277	0.3508	0.592284	0.963366
SVR	0.245733	0.283928	0.532849	0.977686
Rnd. Forest	0.240971	0.281742	0.530794	0.978178

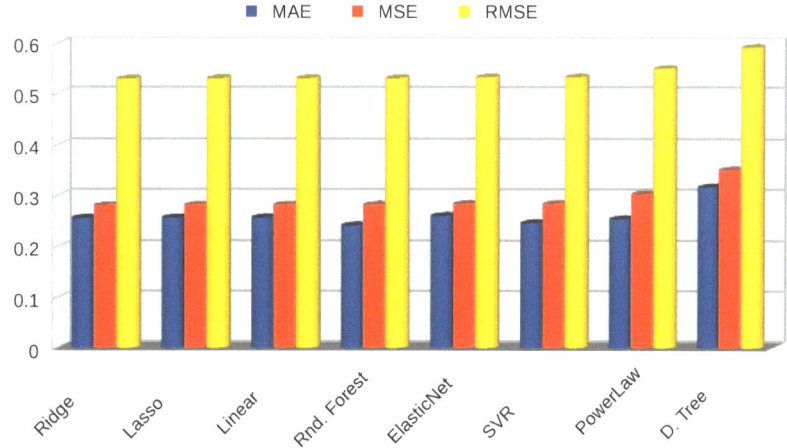

Figure 10. Model errors [m s^{-1}].

Figure 11. Performance scores.

When the models are trained using all the information available from the mesoscale and observed datasets, the best results were obtained by the Random forest, followed by the support vector machine, and now in third place we have again a regression model, but this time not the linear one but the penalized Ridge regression. The RF model had a MAE = 0.24097, MSE = 0.28174, RMSE = 0.53079, R^2 = 0.9781. As expected, all the ML models outperformed the power law method. Figure 12 shows the correlation of WS predicted by the Ridge regression with the recorded by the met mast; Figure 13 shows an example of the Ridge model predicted time series, in a period of high wind speed variability.

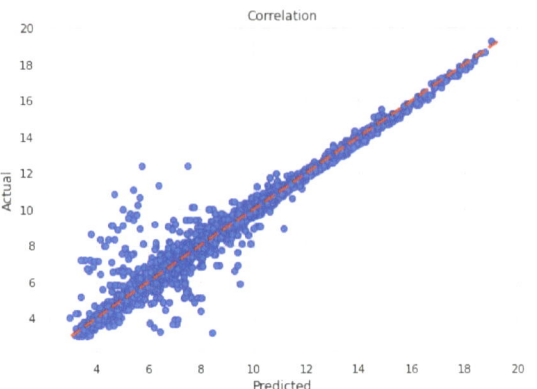

Figure 12. Correlation of the Ridge regression predictions in m s^{-1}.

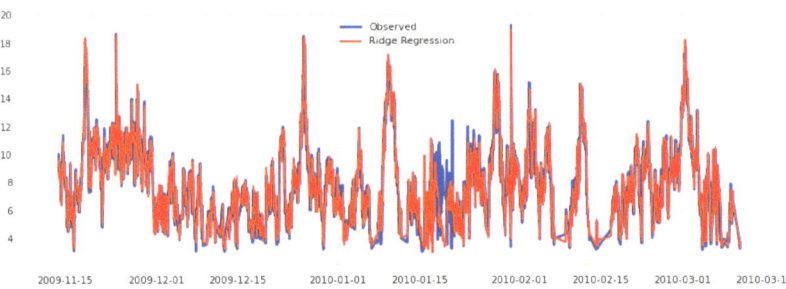

Figure 13. Wind speeds prediction from the Ridge regression using mesoscale and observed data. The units are in m s^{-1}.

4. Discussion

4.1. Scenario A: Extrapolate Wind Speed Using Mesoscale Data from Newa

Based on the R^2, the Decision Tree scored far below its peers. According to the literature, it was expected to obtain better results than the SVR. However, its performance is below all the linear regressions. The regressions models do not extract most features' information; they focus the learning on only wind speeds. For the SVR, the kernel generated a hyperplane that contains above 62% of the attributes transformations. It is verified that the Decision tree cannot predict accurately since its MSE is very high, and only 40% of the variance in the WS 102 m is collectively explained by its regression. However, when many decision trees are combined in a single model, which is known as a Random Forest (which belongs to the ML family of ensemble methods), the results obtained in terms of error and performance are much higher. In fact random forest gets the best score among all the methods. This is due to a wide range of decision paths are generated, which manage to include many more features of the training set, than when using a single decision tree.

After determining that RF was the best method for scenario A, it was compared against the power law method. In a first comparison, the Random forest was trained exclusively with the same mesoscale information that the power law method uses, that is, the wind speed at 50 m and the wind shear coefficient (using the wind speed at 75 m). In this case, the RF obtained an improvement of the R^2 of 33% compared to that obtained by the power law method. For a second comparison, a RF is trained again, but this time using all the information that the NEWA mesoscale model provides, mathematical and statistical transformations carried out in the featuring engineering process (which managed to relate the atmospheric variables, with their effects on wind speed), and an extensive hyperparameter tuning using parallel computing in Pyspark. As expected, by using all

the information available from the NEWA mesoscale model, the R2 of the predictions was further improved in the wind extrapolation. In this case, the RF obtained an R^2 42% better than that of the power law method, and 9% better than the obtained by the RF model used for the first comparison.

We rank the features' contributions for the RF model that was trained with all the information from the NEWA mesoscale model (38 features in total) in Figure 14. The more outstanding ones come from statistical transformations. However, we noticed that the model takes information from all the features, which is the main advantage of the ML techniques over the power law method, where it can only use two wind speeds at different heights.

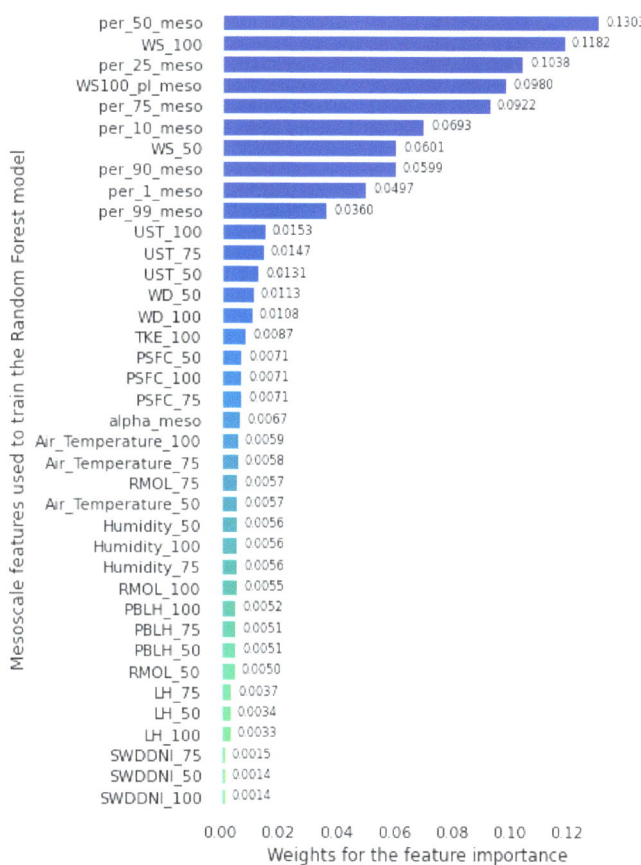

Figure 14. Feature importance for the best model in Scenario A: the random forest. Only mesoscale data for the model training. The sum of all the weights is one.

One advantage of the RF over other ML techniques is that it captures the non-linear features (avoid under-fitting). We can demonstrate this by having a normal distribution of the residuals and the feature utilization rank. The RF relies on a decision making process that accounts for every single possible combination among all the features. However, it explains why the RF model is as well the most expensive in computational terms.

4.2. Scenario B: Extrapolate Wind Speeds Using Data from a Met Mast

We can appreciate that the scores between the ML models do not differ significantly from each other in this scenario. Based on the R^2, the decision tree still scored below its peers. Remarkably, with the observed data, the Linear regressor stands in third place from the best models. According to feature importance for linear regression in this scenario (Figure 15), the model learns directly from the power-law feature (that inherit the information from the wind speeds at 50 and 75 m), and the wind shear to make the predictions. This model also takes statistical features such as the mean and the standard deviation as secondary teachers. The supervisor model tended to take more statistical analysis along with the mathematical transformations. Surprisingly, the best score is obtained when the RF learned almost entirely (98%) from the power-law feature, leaving the remaining 2% among the other features (Figure 16). As a result, the ML model's performance is similar to the traditional method. However, the ML models are still better because they correct some outliers that the power-law does not. The wind speeds predictions for all the models in the range of 4 to 10 m s^{-1} deviate the most from reality. It was expected because the significant concentration of outliers in the data are in that range of high fluctuations that reflect the volatile nature of wind speed.

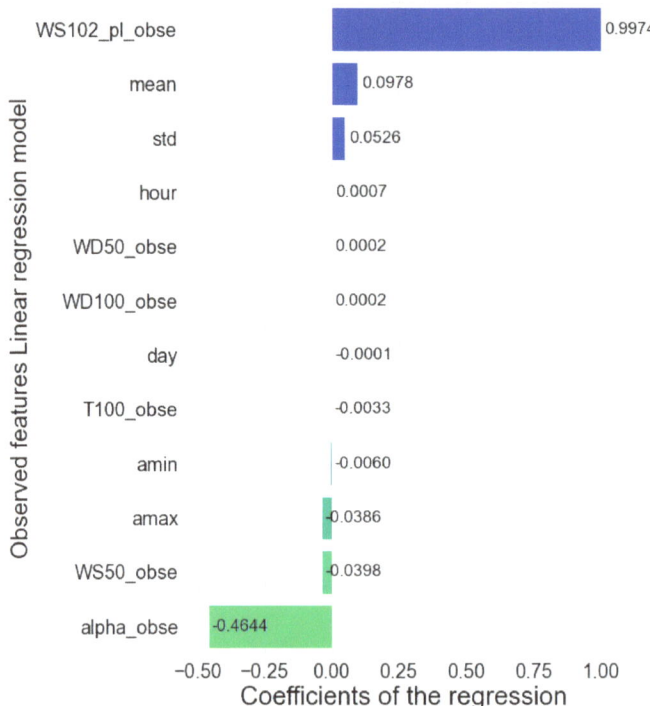

Figure 15. Feature importance for linear regression in Scenario B.

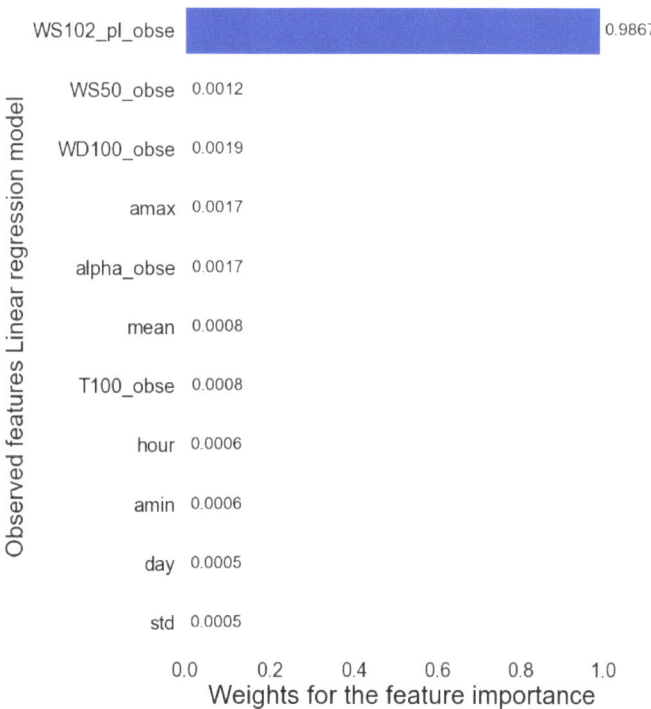

Figure 16. Feature importance for Random Forest in Scenario B.

The Linear Regression model exceeds the quality of the Power Law predictions by 1.8% in the R^2 and is lower than RF by only 0.23%. The error analysis shows that the LR is just as robust as the RF to outliers and data with a high percentage of abrupt change. In computational costs, the RF is way more expensive than the LR. On average, the RF model requires 10 min to solve the task, while LR requires only 5 s. Additionally, the hyperparameter tuning for the RF model took 54 h using distributing computing on three CPUs with six cores each. The LR does not require hyperparameter tuning. Therefore, the recommended model for this scenario is the LR.

4.3. Scenario C: Extrapolate Wind Speeds Using Data from a Met Mast and Mesoscale from the NEWA

The ML models surpass the Power Law method as in the previous two scenarios. Among them, the best model performance belongs to the Random forest, followed by the support vector machine, and in third place, we have a penalized ridge regression. Not all the models benefited from the combination of met mast features and NEWA features, as is the case of Decision Tree, which decreased its R^2 by 0.7%, and Random Forest, which dropped by 0.059%. This can be explained by recognizing that the problem is largely solved with the features derived from the observed data. Additionally, adding more features to train the RF creates more branches, but in the end, the trees decided to go through one of the branches and end up in a leaf with a lower bias for the prediction, omitting most of the other branches. The models that benefited most in this scenario were the Ridge regression that increases its performance on R^2 by 0.19% and Lasso by 0.10%. For Elastic Net, linear regression and SVR, there was neither improvement nor degradation on the scores; they just omitted the NEWA features completely.

Using Grid Search CV in the ElasticNet model, the best value for alpha was found: 0.01. This means that the model was adjusted almost purely in a Ridge configuration penalized

for L2 Norm. The RF scores in R^2, 0.09% higher than the Ridge model, and the SVR only 0.012% more. At this scale, we can conclude that the SVR and the Ridge regressor have the same score. However, if we compare the MSE, the Ridge regressor outperforms all the others models. Comparing computational costs, the Ridge model took 2.28 s to solve the task; meanwhile, the RF and the SVR took almost 2 h. The GridSearch Cross-Validation for the RF and SVM was not carried because it exceeded three days of compilation time using a cluster with three distributed nodes. Based on the previous findings, the task for scenario C has to be solved by a Ridge regression since RF, and SVR models are too computationally expensive. Furthermore, they do not give any considerable improvement in the predictions. The features most studied by the Ridge model (Figure 17) in descending order were: mathematical transformations, power-law, wind shear, mesoscale WS at 100 m, air temperatures, WS observation at 50 m, humidity, Turbulent Kinetic Energy, and Friction Velocities at different heights.

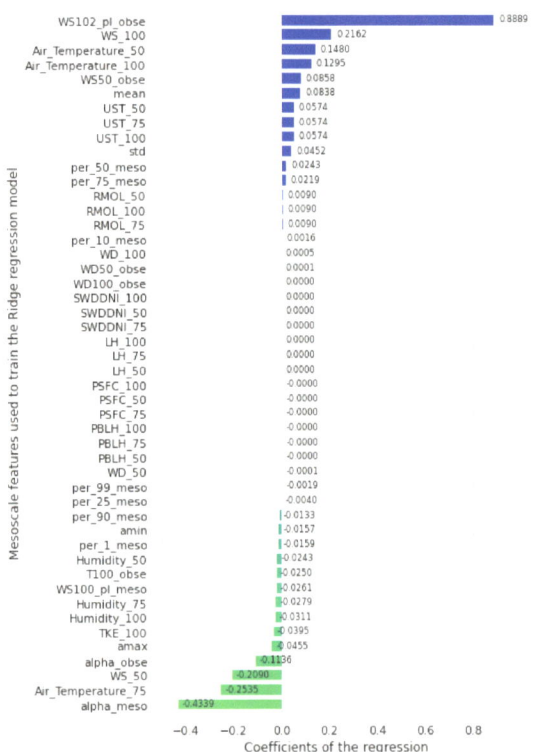

Figure 17. Feature importance for the ridge regression in Scenario C.

5. Conclusions

Using several sources of information (as is the case when using mesoscale data) can lead to an imbalanced data set that is not useful for training an ML model. Therefore, a feature engineering process is required to derive valuable features from the NEWA mesocale and observed data. The mathematical transformation stands out in the feature importance for all three scenarios. The mathematical features link lower wind speeds, temperatures, humidities, and other climate variables well to the higher height objective variable. By training the model with these features, the model does not infer from scratch; on the contrary, it provides a solid base over other assumptions made by the models. A group of 33 features out of 69 are the most suitable to train the models, based on several trial and error experiments and an analysis of the relations between the features and the physical

phenomena that rule the wind speed behavior. The 33 features are: seven come from the measured data, four are focused on temporal attributes, 20 come from the mesoscale model, and two are related to spatial characteristics.

In all the experiments, the machine learning methods trained with mesoscale data, observed data, or a combination of both are superior, as expected, to the power-law method in metrics such as the MAE, MSE, RMSE, and the R^2. When modeled data is available, the predictions become more complex and should be used in an assembled model such as Random Forest (despite its expensive computational demands), which achieves an acceptable prediction. Moreover, when meteorological mast data is available, a regression model at a cheaper computational cost could achieve similar performance to the one conducted by the random forests. Linear regression proved particularly effective when predicting using only measured data, while the Ridge model performed best when mesoscale information is combined with observed data. It was demonstrated that the models depend entirely on how they are trained. The most crucial phase of the machine learning process is to guarantee success in the steps before the modeling phase, such as data cleaning, feature engineering, imputation methods, and construction of the training and evaluation sets. The best scenario result was obtained when mesoscale and observed data were used together. As a result, we obtain better model predictions because the model will use climate information, such as temperatures, humidity, pressures, friction velocities, and wind directions, among other variables.

This work opens possible avenues for new research. On the one hand, it would be essential to determine the level of confidence and uncertainty that a wind energy resource assessment would deliver if a machine learning method is used instead of the power law. More importantly, if an acceptable level of uncertainty is obtained when only modeled data is used. It could lead us in the future to determine if the measurement campaigns can be partially avoided or their costs reduced by not needing to use tall measurement towers. Additionally, an optimized mesoscale model with a coarse resolution could be used to determine if the ML models will assign a higher priority to the wind climate atmospheric conditions, which are, in the end, the direct drivers for the wind velocity, and improve in that way the predictions. Finally, ways to improve the model based on the Random Forest could be investigated, for example, using a high-performance computing platform, which allows a more extensive and precise hyperparameter tuning of the model. Using an HPC platform, one could also propose an ensemble model between the Random Forest and a neural network or a multi-gen genetic programming-based model to catch all the remaining relations between the target variable and the features the RF cannot describe.

Author Contributions: Conceptualization L.B., P.L. and H.T.; methodology, L.B. and H.T.; software, L.B.; validation, L.B.; formal analysis, L.B.; investigation, L.B.; resources, L.B. and P.L.; data curation, L.B.; writing—original draft preparation, L.B.; writing—review and editing, H.T.; visualization, L.B.; supervision, H.T.; project administration, P.L.; funding acquisition, P.L. All authors have read and agreed to the published version of the manuscript.

Funding: This research was funded by and innovation project of the company DNV—Energy Systems, Northern Europe.

Institutional Review Board Statement: Not applicable.

Informed Consent Statement: Not applicable.

Data Availability Statement: Not applicable.

Acknowledgments: To DNV—Energy Systems—Germany, and the University of Oldenburg, for financing and supporting this research. Additionally, a recognition to the people who developed and maintained the software used in the models: Scikit-Learn, PySpark, Pandas, Matplotlib, and Python.

Conflicts of Interest: The authors declare no conflict of interest.

Abbreviations

AI	Artificial Intelligence
ANN	Artificial Neural Network
CRISP-DM	Cross Industrial Standard Process For Data Mining
CV	Cross-Validation
DNN	Deep Neural Network
DT	Decision Trees
LR	Linear Regression
RG	Ridge Regression
MAE	Mean Absolute Error
MAPE	Mean Absolute Percent Error
ML	Machine Learning
MSE	Mean Square Error
NaN	Not a Number Value
NEWA	New European Wind Atlas
OLS	Ordinary Least Squares
PBLH	Planetary Boundary Layer Height
PDF	Probability Density Function
RF	Random Forest
RMSE	Root Mean Square Error
SVM	Support Vector Machine
WPD	Wind Power Density
WRF	Weather Research Forecasting model
WS	Wind Speed

Appendix A. Ml Models Configuration

The following tables present the configuration of the machine learning algorithms that best performed in the cross validation test for each of the three scenarios proposed in Section 2.3.

Scenario A: Hyperparameter Tuning Results	
Model	Model configuration
Linear Regressor	None
Ridge Regressor	Score: 0.627854, Alpha:0.001, Scaler: None
Lasso Regressor	Best parameter (CV score=0.628):{'regressor__alpha': 0.001, 'scaler': StandardScaler()}
Elastic Net Regressor	Best parameter (CV score=0.628): {'regressor__alpha': 0.001, 'scaler': StandardScaler()}
Decision Tree	Best parameter (CV score=0.566): {'regressor__criterion': 'mse', 'regressor__max_depth': 3, 'regressor__max_features': 8, 'regressor__min_samples_leaf': 6, 'scaler': RobustScaler()}
Support Vector Regressor	Best parameter (CV score=0.625): {'kernel': 'rbf', 'gamma': 'auto', 'coef0': 0.01, 'C':0.01, 'epsilon':0.0005, scaler: StandardScaler()}}
Random Forest	Best parameter (CV score=0.647):{'scaler': StandardScaler(), 'regressor__n_estimators': 1200, 'regressor__min_samples_split': 2, 'regressor__min_samples_leaf': 4, 'regressor__max_features': 'sqrt', 'regressor__max_depth': 10, 'regressor__bootstrap': True}

Figure A1. Hyperparameter configuration for the models of Scenario A.

Scenario B: Hyperparameter Tuning Results	
Model	Model configuration
Linear Regressor	None
Ridge Regressor	Alpha Scaler Score 30 1.0 None 0.975386
Lasso Regressor	Best parameter (CV score=0.975): {'regressor__alpha': 0.0001, 'scaler': RobustScaler()}
Elastic Net Regressor	Best parameter (CV score=0.975): {'regressor__alpha': 0.0001, 'scaler': StandardScaler()}
Decision Tree	Best parameter (CV score=0.971): {'regressor__criterion': 'mae', 'regressor__max_depth': None, 'regressor__max_features': 6, 'regressor__min_samples_leaf': 7, 'scaler': RobustScaler()}
Support Vector Regressor	Best parameter (CV score=0.625): {'kernel': 'poly', 'gamma': 'auto', 'coef0': 0.001, 'C':0.1, 'epsilon':0.01, scaler': StandardScaler()}}
Random Forest	Best parameter (CV score=0.979): {'scaler': MinMaxScaler(), 'regressor__n_estimators': 800, 'regressor__min_samples_split': 2, 'regressor__min_samples_leaf': 4, 'regressor__max_features': 'auto', 'regressor__max_depth': 100, 'regressor__bootstrap': True}

Figure A2. Hyperparameter configuration for the models of Scenario B.

Scenario C: Hyperparameter Tuning Results	
Model	Model configuration
Linear Regressor	None
Ridge Regressor	Best parameter (CV score=0.976): {'regressor__alpha': 5, 'scaler': None}
Lasso Regressor	Best parameter (CV score=0.975): {'regressor__alpha': 0.0001, 'scaler': RobustScaler()}
Elastic Net Regressor	Best parameter (CV score=0.975): {'regressor__alpha': 0.0001, 'scaler': StandardScaler()}Best parameter (CV
Decision Tree	Best parameter (CV score=0.909): {'regressor__criterion': 'mae', 'regressor__max_depth': None, 'regressor__max_features': 7, 'regressor__min_samples_leaf': 2, 'scaler': StandardScaler()}
Support Vector Regressor	Best parameter (CV score=0.625): {'kernel': 'linear', 'gamma': 'auto', 'coef0': 0.001, 'C':0.1, 'epsilon':0.01, scaler': MinMaxScaler()}}
Random Forest	Best parameter (CV score=0.978): {'scaler': StandardScaler(), 'regressor__n_estimators': 1800, 'regressor__min_samples_split': 5, 'regressor__min_samples_leaf': 4, 'regressor__max_features': 'auto', 'regressor__max_depth': 90, 'regressor__bootstrap': True}

Figure A3. Hyperparameter configuration for the models of Scenario C.

Appendix B. Performance Metrics

The indicators that can be used to compare model performance in regression tasks are presented to continue. For the final evaluation of the R^2, a cross-validation with 5 folds was used. There, N represents the total number of data points, y the actual value, $\hat{y}$ the predicted value, $y - \hat{y}$ is the residual, and $\bar{y}$ is the mean of the observed data denoted by $\bar{y} = \sum_{i=1}^{N} y_i$.

Mean Absolute Error (MAE): It asses the absolute differences, and is less sensitive to outliers. Thus, it is good for comparing different models.

$$MAE = \frac{1}{N} \sum_{i=1}^{N} |y - \hat{y}| \quad \text{(A1)}$$

Mean squared error (*MSE*): It is the average of the squared differences. Even the small errors are penalized. It can lead to an over-estimation of how bad the model is. MSE is used to determine the extent to which the model fits the data.

$$MSE = \frac{1}{N}\sum_{i=1}^{N}(y-\hat{y})^2 \tag{A2}$$

Root Mean Squared Error (*RMSE*): It is the standard deviation of the residuals. It has a high penalty on large errors (the errors are first squared before averaging); thus, it is used when avoiding bigger errors is desirable.

$$RMSE = \sqrt{\frac{\sum_{i=1}^{N}(y-\hat{y})^2}{N}} \tag{A3}$$

Coefficient of determination (R^2): It compares the regression model with a constant baseline and tells us how much our model differs from the original one. It is related to the correlation coefficient, r, which tells you how strong of a linear relationship there is between two variables.

$$R^2 = 1 - \frac{MSE(Model)}{MSE(Baseline)}$$
$$R^2 = 1 - \frac{\sum_{i=1}^{N}(y-\hat{y})^2}{\sum_{i=1}^{N}(y-\bar{y})^2} \tag{A4}$$

References

1. International Renewable Energy Agency. *Capacity Statistics 2019*; Technical Report; IRENA: Abu Dhabi, United Arab Emirates, 2020.
2. International Renewable Energy Agency. *Future of Wind 2019*; Technical Report; IRENA: Abu Dhabi, United Arab Emirates, 2020.
3. Roeth, J. *Wind Resource Assessment Handbook*; AWS Truepower: New York, NY, USA, 2010.
4. Gasch, R.; Twele, J. (Eds.) *Wind Power Plants: Fundamentals, Design, Construction and Operation*; Springer: Berlin/Heidelberg, Germany, 2012. [CrossRef]
5. Valsaraj, P.; Drisya, G.V.; Kumar, K.S. A Novel Approach for the Extrapolation of Wind Speed Time Series to Higher Altitude Using Machine Learning Model. In Proceedings of the 2018 International CET Conference on Control, Communication, and Computing (IC4), Thiruvananthapuram, India, 5–7 July 2018; pp. 112–115. [CrossRef]
6. Mohandes, M.A.; Rehman, S. Wind Speed Extrapolation Using Machine Learning Methods and LiDAR Measurements. *IEEE Access* 2018, *6*, 77634–77642. [CrossRef]
7. Europe Union. *The New European Wind Atlas (NEWA)*; Geoscientific Model Development: Munich, Germany, 2020.
8. Manwell, J.F. *Wind Energy Explained: Theory, Design and Application*; Wiley: Hoboken, NJ, USA, 2009.
9. Eyecioglu, O.; Hangun, B.; Kayisli, K.; Yesilbudak, M. Performance comparison of different machine learning algorithms on the prediction of wind turbine power generation. In Proceedings of the 8th International Conference on Renewable Energy Research and Applications, ICRERA 2019, Brasov, Romania, 3–6 November 2019; pp. 922–926. [CrossRef]
10. Malakhov, A.; Goncharov, F. Testing proaches for wind plants power output. In Proceedings of the 2019 International Youth Conference on Radio Electronics, Electrical and Power Engineering (REEPE), Moscow, Russia, 14–15 March 2019. [CrossRef]
11. Liu, Y.; Zhang, H. An Empirical Study on Machine Learning Models for Wind Power Predictions. In Proceedings of the 2016 15th IEEE International Conference on Machine Learning and Applications (ICMLA), Anaheim, CA, USA, 18–20 December 2017; pp. 758–763. [CrossRef]
12. Ji, G.R.; Han, P.; Zhai, Y.J. Wind speed forecasting based on support vector machine with forecasting error estimation. In Proceedings of the Sixth International Conference on Machine Learning and Cybernetics, ICMLC 2007, Hong Kong, China, 19–22 August 2007; Volume 5, pp. 2735–2739. [CrossRef]
13. Brahimi, T. Using artificial intelligence to predict wind speed for energy application in Saudi Arabia. *Energies* 2019, *12*, 4669. [CrossRef]
14. Liu, Y.; Wang, J.; Collett, I.; Morton, Y.J. A Machine Learning Framework for Real Data Gnss-R Wind Speed Retrieval. In Proceedings of the IGARSS 2019—2019 IEEE International Geoscience and Remote Sensing Symposium, Yokohama, Japan, 28 July–2 August 2019; pp. 8707–8710. [CrossRef]
15. Ali, M.E.K.; Hassan, M.Z.; Ali, A.B.; Kumar, J. Prediction of Wind Speed Using Real Data: An Analysis of Statistical Machine Learning Techniques. In Proceedings of the 2017 4th Asia-Pacific World Congress on Computer Science and Engineering, APWC on CSE 2017, Mana Island, Fiji, 11–13 December 2017; pp. 259–264. [CrossRef]

16. Ak, R.; Fink, O.; Zio, E. Two Machine Learning Approaches for Short-Term Wind Speed Time-Series Prediction. *IEEE Trans. Neural Netw. Learn. Syst.* **2016**, *27*, 1734–1747. [CrossRef] [PubMed]
17. Albrecht, C.; Klesitz, M. Long-term correlation of wind measurements using neural networks: A new method for post-processing short-time measurement data. In Proceedings of the European Wind Energy Conference and Exhibition 2007, EWEC 2007, Milan, Italy, 7–10 May 2007; Volume 2, pp. 755–762.
18. Cheggaga, N.; Ettoumi, F.Y. A Neural Network Solution for Extrapolation of Wind Speeds at Heights Ranging for Improving the Estimation of Wind Producible. *Wind. Eng.* **2011**, *35*, 33–53. [CrossRef]
19. Türkan, Y.S.; Yumurtacı Aydoğmuş, H.; Erdal, H. The prediction of the wind speed at different heights by machine learning methods. *Int. J. Optim. Control. Theor. Appl.* **2016**, *6*, 179. [CrossRef]
20. Cheggaga, N. A new artificial neural network–power law model for vertical wind speed extrapolation for improving wind resource assessment. *Wind. Eng.* **2018**, *42*, 510–522. [CrossRef]
21. Vassallo, D.; Krishnamurthy, R.; Fernando, H.J.S. Decreasing wind speed extrapolation error via domain-specific feature extraction and selection. *Wind. Energy Sci.* **2020**, *5*, 959–975. [CrossRef]
22. Bodini, N.; Optis, M. How accurate is a machine learning-based wind speed extrapolation under a round-robin approach? *J. Phys. Conf. Ser.* **2020**, *1618*, 062037. [CrossRef]
23. Adli, F.; Cheggaga, N.; Farouk, H. Vertical wind speed extrapolation: Modelling using a response surface methodology (RSM) based on unconventional designs. *Wind. Eng.* **2022**, *35*, 33–53. [CrossRef]
24. Emeksiz, C. Multi-gen genetic programming based improved innovative model for extrapolation of wind data at high altitudes, case study: Turkey. *Comput. Electr. Eng.* **2022**, *100*, 107966. [CrossRef]
25. Shi, T.; He, G.; Mu, Y. Random Forest Algorithm Based on Genetic Algorithm Optimization for Property-Related Crime Prediction. In *Proceedings of the 2019 International Conference on Computer, Network, Communication and Information Systems (CNCI 2019), Qingdao, China, 27–29 March 2019*; Atlantis Press: Amsterdam, The Netherlands, 2019; pp. 526–531. [CrossRef]
26. Farooq, F.; Nasir Amin, M.; Khan, K.; Rehan Sadiq, M.; Faisal Javed, M.; Aslam, F.; Alyousef, R. A Comparative Study of Random Forest and Genetic Engineering Programming for the Prediction of Compressive Strength of High Strength Concrete (HSC). *Appl. Sci.* **2020**, *10*, 7330. [CrossRef]
27. Feng, J. Artificial Intelligence for Wind Energy. In *A State of the Art Report 978-87-93549-48-7*; DTU Wind Energy: Lyngby, Denmark, 2019.
28. Chapman, P.; Clinton, J.; Kerber, R.Y.; Shearer, C.; Khabaza, T. *CRISP-DM 1.0. Step-by-Step Data Mining Guide*; SPSS Inc.: Chicago, IL, USA, 2000.
29. Subasi, A. *Practical Machine Learning for Data Analysis Using Python*; Elsevier: Amsterdam, The Netherlands, 2020. [CrossRef]
30. Baquero, L. *Extraction of Section 2.5 from the Thesis Book: Theory about Machine Learning Used in This Research*; ResearchGate: Berlin, Germany, 2022; Volume 1, pp. 25–36.
31. Zaharia, M.; Xin, R.S.; Wendell, P.; Das, T.; Armbrust, M.; Dave, A.; Meng, X.; Rosen, J.; Venkataraman, S.; Franklin, M.J.; et al. Apache spark: A unified engine for big data processing. *Commun. ACM* **2016**, *59*, 56–65. [CrossRef]
32. Pedregosa, F.; Varoquaux, G.; Gramfort, A.; Michel, V.; Thirion, B.; Grisel, O.; Blondel, M.; Prettenhofer, P.; Weiss, R.; Dubourg, V.; et al. Scikit-learn: Machine learning in Python. *J. Mach. Learn. Res.* **2011**, *12*, 2825–2830.
33. Seabold, S.; Perktold, J. Statsmodels: Econometric and statistical modeling with python. In Proceedings of the 9th Python in Science Conference, Austin, TX, USA, 28 June–3 July 2010.
34. McKinney, W. Data structures for statistical computing in python. In Proceedings of the 9th Python in Science Conference, Austin, TX, USA, 28 June–3 July 2010; Volume 445, pp. 51–56.
35. Hahmann, A.N.; Sile, T.; Witha, B.; Davis, N.N.; Dörenkämper, M.; Ezber, Y.; García-Bustamante, E.; González Rouco, J.F.; Navarro, J.; Olsen, B.T.; et al. *The Making of the New European Wind Atlas, Part 1: Model Sensitivity*; Technical Report; Model Sensitivity; Geoscientific Model Development: Munich, Germany, 2020. [CrossRef]
36. Camuffo, D. *Microclimate for Cultural Heritage: Conservation, Restoration, and Maintenance of Indoor and Outdoor Monuments*, 2nd ed.; Elsevier: Amsterdam, The Netherland; Boston, MA, USA, 2014.
37. DNV GL. *Framework for Assurance of DataDriven Algorithms and Models*; DNV GL: Oslo, Norway, 2020.

Article

Numerical Investigation of Multi-Floater Truss-Type Wave Energy Convertor Platform

Ruijia Jin *, Jiawei Wang, Hanbao Chen, Baolei Geng and Zhen Liu *

National Engineering Research Center of Port Hydraulic Construction Technology, Tianjin Research Institute for Water Transport Engineering M.O.T., Tianjin 300456, China
* Correspondence: ruijia_jin@163.com (R.J.); wave_2006@126.com (Z.L.)

Abstract: In order to solve the hydrodynamic characteristics of the multi-floater truss-type wave energy convertor (WEC) platform, the mathematical model is established by using the high-order boundary element method based on potential flow theory, in which the floater and the platform are connected by the floating arm based on the lever principle. The mathematical model is applied to study the heave motion response of each floater of the multi-floater truss-type WEC platform, and the effects of the floater number and the floater arrangement on the motion responses of floaters, as well as the power generation of the WEC platform are analyzed. The effect of the hydraulic cylinder on the floater is simulated by linear damping, and then, the work of the hydraulic cylinder is used to generate electricity, so as to achieve the purpose of simulating the multi-floater WEC power generation device. Some useful conclusions are obtained through calculation, which can provide data support for the corresponding platform.

Keywords: multi-floater WEC platform; potential flow theory; numerical simulation; wave energy

Citation: Jin, R.; Wang, J.; Chen, H.; Geng, B.; Liu, Z. Numerical Investigation of Multi-Floater Truss-Type Wave Energy Convertor Platform. *Energies* **2022**, *15*, 5675. https://doi.org/10.3390/en15155675

Academic Editor: Hua Li

Received: 23 June 2022
Accepted: 1 August 2022
Published: 4 August 2022

Publisher's Note: MDPI stays neutral with regard to jurisdictional claims in published maps and institutional affiliations.

Copyright: © 2022 by the authors. Licensee MDPI, Basel, Switzerland. This article is an open access article distributed under the terms and conditions of the Creative Commons Attribution (CC BY) license (https://creativecommons.org/licenses/by/4.0/).

1. Introduction

To cope with the great challenges brought about by the energy crisis, it has become an urgent task for countries all over the world to explore clean and renewable energy. As a new clean energy, compared with other renewable energy, wave energy has obvious advantages. Firstly, the energy utilization efficiency of wave energy is relatively high, and it has the highest energy density among renewable energy. The process of developing and utilizing wave energy has little impact on the environment and aquatic organisms. In addition, the wave energy power generation device can operate 90% of the time, while the operation of wind energy utilization equipment and solar energy utilization equipment is greatly affected by the environment, and the normal working time is only one third of the wave energy. Zhang et al. [1] reviewed the progress in wave energy technologies in China briefly, proposed the development direction and prospect in the future, and hopes for international cooperation to establish the market and production facilities and to share experiences.

Wave energy technology has experienced a hundred years of development and now has many practical applications. For the oscillating floater-type wave energy convertor (WEC), prototypes have been put into use in many countries. Norway developed a buoy-type oscillating float device [2]. The L9 oscillating floater device developed in Sweden uses a flat cylinder float as an energy absorber. The generator used in the device was fixed to the seabed. The floater was connected to the generator with a tensioned cable, and a linear motor was used as an energy conversion device [3]. In September 2008, the United States successfully conducted a sea trial of the L-10 oscillating float wave energy power generation device in Newport, Oregon. Its power output system adopts a linear motor with a rated installed capacity of 10 kW [4]. Carnegie wave energy of Australia developed a "CETO" wave energy device, which used a large underwater float to connect with a turbine pump set installed on the seabed for power generation [5].

Many scholars have been carrying out continuous research on the oscillating floater WEC device. The analytical method, numerical simulation method, and physical model are the most common research methods. For the analytical method, Korde calculated the coupled vibration system based on the analytical method under regular waves, studied the floater connected by the elastic damping device and the mass disc in the water, and proposed the concept of transforming the system into two degrees of freedom to adapt to different wave spectra [6]. Caska and Finnigan used the analytical method to study the hydrodynamic performance of a wave energy device with a cylindrical floater articulated at the bottom of the sea, and the conclusion was drawn that the nonlinear drag force has a great influence on the motion performance of the float [7]. Zhao et al. developed an analytical model based on linear potential flow theory and matching eigenfunction expansion technique to investigate the hydrodynamics of a two-dimensional breakwater with an oscillating buoy wave energy converter [8].

Due to the complex shape of the oscillating floater, more and more scholars still use numerical simulation to carry out their research. Cao applied the software AQWA to study the influence of the radius and mass on the hydrodynamic coefficient of the cylindrical floater systematically, analyzed the hydrodynamic performance of the conical bottom floater with different angles, and compared the hydrodynamic coefficients of the floater [9]. Zhou established the motion model of the oscillating floater wave energy generator using OrcaFlex, studied the nonlinear hydrodynamic characteristics of the device, compared the vibration efficiency of the floater under different masses and electromagnetic damping coefficients, and calculated the corresponding power generation according to the basic formula [10]. Zhou built the hydrodynamic model of the pitching floater based on AQWA, studied the parameterized shape scheme of the three common shapes of the floater, hemisphere bottom, cone bottom, and platform bottom, and analyzed the hydrodynamic characteristics of the pitching floater moving in waves [11]. Zhang et al. investigated the hydrodynamic performance of a dual-floater hybrid system consisting of a floating breakwater and an oscillating-buoy-type wave energy converter using Star-CCM+ Computational Fluid Dynamics software, and the research made wave energy economically competitive and commercial-scale wave power operations possible [12]. Finnegan et al. developed a computational fluid dynamics model of the CECO wave energy converter (WEC) using the commercial software ANSYS CFX. The numerical model was used to investigate the nonlinear effects on the motions of CECO and to obtain more insights in relation to wave loading during a wave cycle and the viscous effects associated with the dissipation of energy in the flow around its floaters [13]. Luan et al. established a three-dimensional numerical wave tank by STAR-CCM+ and simulated a truncated column in regular waves, and the relationship between the optimal damping constants and wave number was studied [14].

At present, more and more scholars are also focusing on the research of complex nonlinear problems such as the multi-floater array and double-floater device coupling motion. Liu used linear potential flow theory to optimize the floater array design under the condition of regular waves and irregular waves and calculated its motion response and wave energy capture efficiency. The results showed that the array device can make full use of wave energy resources [15]. Yang et al. used a numerical model to study the hydrodynamic response characteristics of float arrays arranged in circumferential, double-column, and single-column directions with different wave directions and the influence of float spacing on the wave energy absorption of each floater [16]. Chandrasekaran and Sricharan deliberated on the numerical analysis of a new, bean-shaped, multi-body floating wave energy converter using an open-source time-domain modeling tool. The authors proposed three different layouts with multiple floats to study the influence of the float number on the device's overall performance [17]. Marchesi et al. developed a numerical model of the energy double system on the basis of the existent laboratory model and simulated new cases of different values of PTO damping and random waves [18]. He et al. established a numerical model to investigate multi-body hydrodynamic interaction

between an octagonal platform and absorber-type wave energy converters and selected a final design. They demonstrated that the multi-body interaction has a remarkable influence on the absorption power [19].

With the improvement of physical model test simulation technology, the complex marine structure of the oscillating floater wave energy convertor device can also be simulated at a small scale in a physical model test. Negri and Malavasi tested physical models of the two systems in a wave flume, which were tested with monochromatic waves [20]. Ramadan et al. conducted an experimental analysis of an enhanced design of a float with an inverted cup for wave energy conversion. The results indicated that the captured efficiency for the float of 30 cm in diameter with a baffle is 19 percent instead of 6 percent for the float of 50 cm in diameter without a baffle. The efficiency was increased three-times more than the conventional design, as well as superior performance under the effect of regular wave patterns was obtained [21]. For the wave energy converter M4, Moreno and Stansby undertook a physical model test for the six-float wave energy converter M4 at a 1:50 scale. They presented the results for angular motion at the PTOs and mooring forces. Wave conditions with different spectral peakedness and multi-directional spreading were applied and energy yield with electricity cost estimated made by 11 offshore sites [22]. Then, they investigated a multi-body linear diffraction–radiation model for the wave energy converter M4, including mean second-order forces and radiation damping, as well as mean excitation force. According to the comparisons of the experimental results, the authors found that the linear modeling gave a reasonable prediction of the response in all wave conditions and power capture when operational, but resulting second-order mean forces only give approximate predictions[23]. Santo et al. analyzed the performance of the M4 wave energy converter off Albany on the south coast of western Australia, an area well-known for almost continuous exposure to long-period swells [24].

In this paper, the potential flow theory is used to establish a numerical model of the interaction between waves and the oscillating-floater-type truss wave energy power-generation platform. The hydrodynamic characteristics of the oscillating floater wave power generation platform under wave action were studied in the time domain. The effects of the floater spacing and the number of floaters on the movement were analyzed. By setting the damping coefficient to simulate the effect of the hydraulic cylinder on the floater movement, the optimal power generation damping of the floater was found. The relevant calculation results can provide a data reference for the design of an oscillating-floater-type truss wave energy power generation platform with a similar shape.

2. Mathematical Formulation

2.1. Calculation of Platform and Floaters

Diffraction theory was adopted for the wave–structure interaction study. A right-hand Cartesian coordinate was established in the computation. One is a space-fixed coordinate system $Oxyz$ with its origin at the still water surface, in which x and y are measured horizontally and z vertically upward. For each floater, a body-fixed coordinate system $O'_i x'_i y'_i z'_i$ was established to describe the motion of each floater (i denotes the ith floater). The sketch is shown in Figure 1.

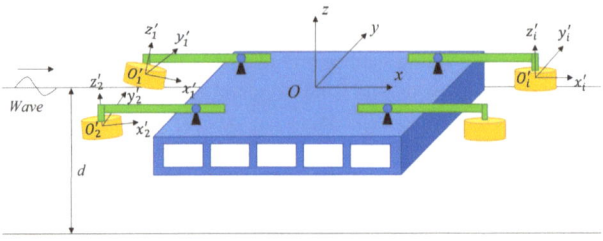

Figure 1. Sketch of the wave and multi-floater truss-type WEC platform.

Under the assumption of ideal fluid, there exists a velocity potential ϕ that satisfies the Laplace equation and boundary conditions within the fluid domain. The velocity potential and wave elevation can be divided into the incident part and scattered part:

$$\phi = \phi_i + \phi_s \tag{1}$$

$$\eta = \eta_i + \eta_s \tag{2}$$

Thus, the scattered potential satisfies the Laplace equation in the domain as follows:

$$\nabla^2 \phi_s = 0 \tag{3}$$

It is subject to the seabed boundary and free surface conditions:

$$\frac{\partial \phi_s}{\partial z} = 0, z = -d \tag{4}$$

$$\begin{cases} \frac{\partial \eta_s}{\partial t} = \frac{\partial \phi_s}{\partial z} \\ \frac{\partial \phi_s}{\partial t} = -g\eta_s \end{cases}, z = 0 \tag{5}$$

For the platform and the floater, the body boundary condition is

$$\frac{\partial \phi_s^i}{\partial \vec{n}} = -\frac{\partial \phi_i^i}{\partial \vec{n}} + \left(\vec{\zeta}^i + \vec{\alpha}^i \times \left(\vec{x}^i - \vec{x}_0^i\right)\right) \cdot \vec{n} \quad (i = 0, 1, 2, \cdots, N) \tag{6}$$

where the superscript i indicates different floaters, $i = 0$ denotes the platform, $i = 1, 2, \cdots, N$ denotes the floater, $\vec{n}$ denotes the unit normal vector, pointing out of the fluid, and $\vec{\zeta}^i = \left(\zeta_1^i, \zeta_2^i, \zeta_3^i\right)$ and $\vec{\alpha}^i = \left(\alpha_1^i, \alpha_2^i, \alpha_3^i\right) = \left(\zeta_4^i, \zeta_5^i, \zeta_6^i\right)$ denote the translation and rotation motion, respectively. $\vec{x}_0^i$ denotes the rotation center.

To numerically solve the boundary value problem, we employ a Rankine source and its image about the seabed as Green's function. The second theorem of Green is applied to the scattered potential and Green's function, and thus, the above boundary value problem is converted to the boundary integral equation.

Once the velocity potential on the body surface is obtained, the wave forces on a body can be computed by integrating the fluid pressure over the mean body surface. The exciting force is expressed as

$$\vec{F}^i = -\rho \int_{S_b} \phi_t \vec{n} ds \quad (i = 0, 1, 2, \cdots, N) \tag{7}$$

where ρ denotes the fluid density. Similarly, $\vec{F}^0$ indicates the wave exciting force on the platform; $\vec{F}^i$ indicates the wave exciting force on the ith floater.

2.2. Calculation of the Truss Structure

The wave forces on the truss structure are obtained by the Morison formula. The wave force on the unit height is

$$f_s = f_i + f_d \tag{8}$$

where f_i indicates the inertia force, whose form is the same as the solution of the nonviscous fluid based on wave theory, and f_d represents the velocity force, whose form is similar to the resistance on the body surface in steady flow. The formulas for the forces acting on the vertical truss structure are

$$\vec{f}_i = C_m \rho \frac{\pi D^2}{4} \vec{a} \tag{9}$$

$$\vec{f}_d = C_d \frac{\rho}{2} D \vec{u} |\vec{u}| \tag{10}$$

where C_m and C_d are the coefficients of inertia and velocity force, respectively; D is the cylinder diameter; $\vec{u}$ and $\vec{a}$ are the velocity and acceleration of a water particle, respectively.

Due to the effect of the platform and floaters on the wave field, the velocity and acceleration of water are composed of incident and scattered waves, respectively. The parts induced by incident waves can be easily obtained through an analytical expression; however, the parts induced by scattered waves should be solved by wave diffraction theory. The velocity of a water particle produced by diffraction potential can be obtained by the integral equation and is given as

$$u_{s,x} = \frac{\partial \phi_{s,x}(\vec{x}_0)}{\partial t} = \int_S \left[\frac{\partial^2 G(\vec{x}, \vec{x}_0)}{\partial \vec{n} \partial x_0} \phi_s(\vec{x}) - \frac{\partial \phi_s(\vec{x})}{\partial \vec{n}} \frac{\partial G(\vec{x}, \vec{x}_0)}{\partial x_0} \right] ds \quad (11)$$

Then, the acceleration of a water particle can be calculated by the time difference:

$$a_{s,x} = \frac{\partial u_{s,x}}{\partial t} = \frac{(u_{s,x})^t - (u_{s,x})^{t-1}}{\Delta t} \quad (12)$$

Because the body has reciprocating motion under the wave action, the velocity and acceleration can be calculated by solving the body motion equation, which are defined as $\vec{u}_b$ and $\vec{a}_b$. Therefore, the relative velocity and acceleration can be written as $\vec{u}_r = \vec{u} - \vec{u}_b$ and $\vec{a}_r = \vec{a} - \vec{a}_b$, respectively. Therefore, the wave force on a vertical truss structure for a unit of height can be written as follows:

$$\vec{f}_i = C_m \rho \frac{\pi D^2}{4} \vec{a}_r \quad (13)$$

$$\vec{f}_d = C_d \frac{\rho}{2} D \vec{u}_r |\vec{u}_r| \quad (14)$$

The validation of the wave force calculation about the truss structure in the large structure was completed in previous research [25].

2.3. Motion Response

The platform and floater are connected by a floating arm and hydraulic cylinder, as shown in Figure 2.

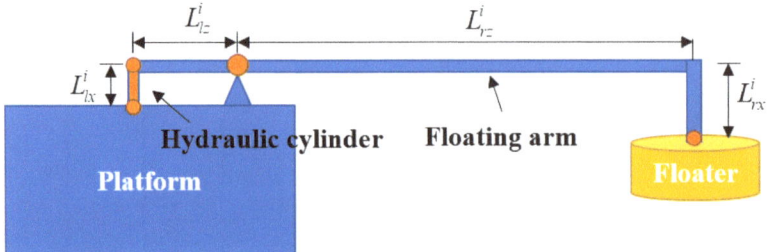

Figure 2. Sketch of the connection relationship of the platform, hydraulic cylinder, and floater.

Based on the lever principle, when the floater is subjected to small wave force, it can push the rear hydraulic cylinder to work and generate power. Therefore, the wave force on the floaters is amplified by the floating arm and acts on the platform, thus affecting the platform movement. The motion equation is required for each floater as follows:

$$[M^i]\{\ddot{\vec{\xi}}^i\} + [B^i]\{\dot{\vec{\xi}}^i\} + [C^i]\{\vec{\xi}^i\} = \{\vec{F}^i\} \quad (15)$$

where $[M^i]$, $[B^i]$ and $[C^i]$ are the mass matrix, damping matrix, and stiffness matrix for each floater. For the platform, the motion equation of the platform is expressed by

$$[M^0]\{\ddot{\vec{\xi}^0}\} + [B^0]\{\dot{\vec{\xi}^0}\} + [C^0]\{\vec{\xi}^0\} = \{\vec{F}^0\} + \sum_{i=1}^{N} \frac{L_r^i}{L_l^i}\{\vec{F}^i\} + \{\vec{F}_T\} \qquad (16)$$

where $[M^0]$, $[B^0]$ and $[C^0]$ are the mass matrix, damping matrix, and stiffness matrix for the platform; L_r^i and L_l^i are the floating arm length of the floater side and platform side, respectively, and $\vec{F}_T$ indicates the Morison force on the truss structure.

3. Model Validation

The study of the waves and WEC platform is actually a multi-floating body interaction problem. To validate the numerical model, the linear wave interaction of a twin-box structure is modeled, as shown in Figure 3. L, B, T, and W in Figure 3 represent the length, width, draft, and spacing of the twin-boxes, respectively, and d represents the water depth. The detailed dimensions and calculation parameters of the square box are shown in Table 1. The mass center of the square box is located 2.56 m directly above the center of the box bottom.

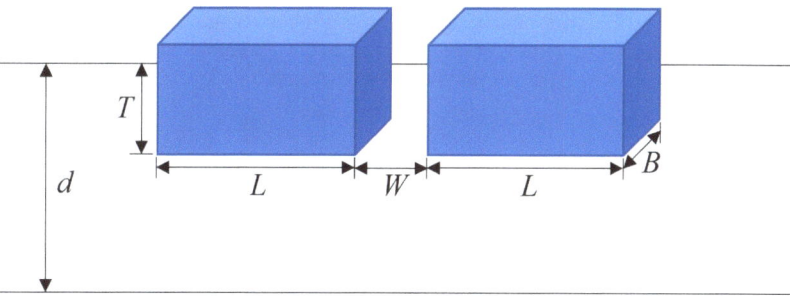

Figure 3. Sketch of the wave interaction a twin-box structure.

Table 1. Parameters of the double-box model.

Symbol	Meaning	Value
L	Length (m)	30
B	Width (m)	22
T	Draft (m)	1.5
d	Water depth (m)	15
W	Distance of twin-boxes (m)	8
R_x	Rotation radius around x-axis (m)	9.0
R_y	Rotation radius around y-axis (m)	6.6
R_z	Rotation radius around z-axis (m)	10.8

This model is applied to calculate the hydrodynamic coefficient and motion response of two free square boxes in regular waves with an incidence angle of 0 degrees. Figure 4 shows the relationship between the wave force and the incident wave frequency of the up-wave box and the back-wave box, respectively. Figure 5 shows the variation of the motion response in the pitch motion of the up-wave box and the back-wave box with the wave frequency.

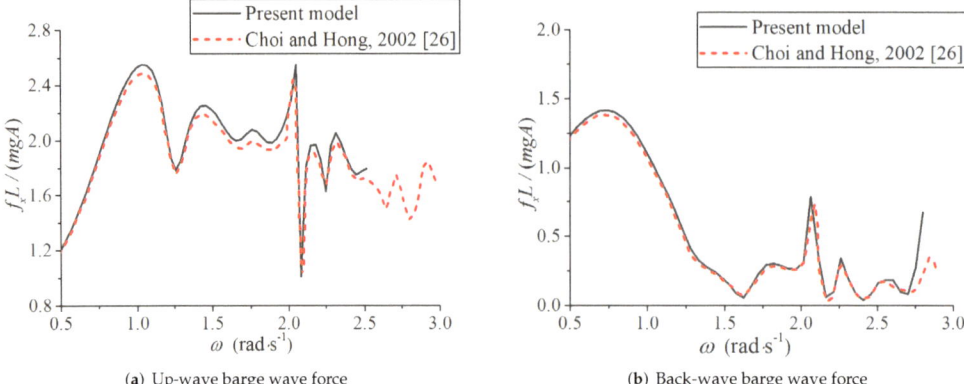

Figure 4. Wave forces of barges in the surge direction: (**a**) up-wave barge; (**b**) back-wave barge [26].

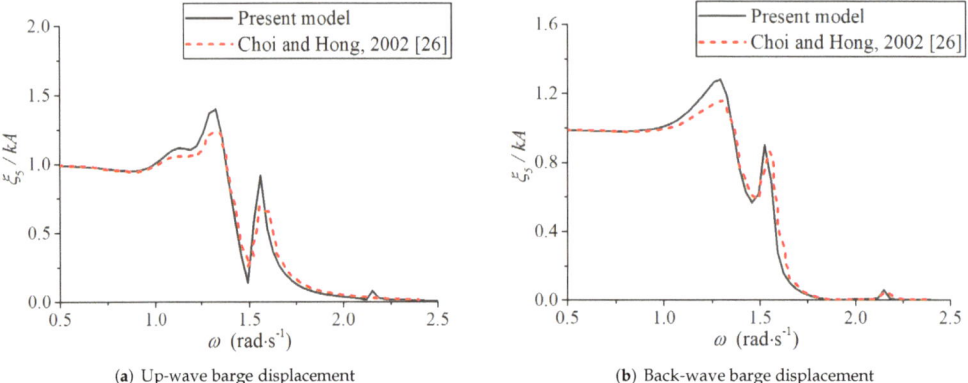

Figure 5. Displacement of barges in the surge direction: (**a**) up-wave barge; (**b**) back-wave barge [26].

In the figure, the wave force and motion response are dimensionless. m, g, L, and A are the mass, gravity acceleration, box length, and wave amplitude, respectively. The results show that the calculation results of the present model are consistent with Choi and Hong (2002) [26]. The calculation results are in good agreement only with some differences in individual frequencies. Therefore, this model can be employed in the subsequent hydrodynamic response calculation of the multi-floater truss-type WEC platform.

4. Results and Discussions

The truss-type WEC platform is composed of oscillating floaters and a platform, which are connected by a floating arm and a hydraulic cylinder in the form of a lever. It is described in Figures 1 and 2 above. The parameters of the floater and platform are shown in Tables 2 and 3. In the platform, there are five large damping boxes in the water to provide buoyancy force. The draft of a damping box is 1.8 m, and the center location is in Table 4.

Table 2. Parameters of the platform.

Meaning	Value
Size of platform	20 m
Mass of platform	1.025×10^5 kg
Mass center of platform	(0.0, 0.0, 0.0)
Rotation center of platform	(0.0, 0.0, 0.0)
Platform moment of inertia I_{xx}	6.56×10^6 kg·m²
Platform moment of inertia I_{yy}	6.56×10^6 kg·m²
Platform moment of inertia I_{zz}	9.72×10^6 kg·m²
Distance between floater and platform edge	5 m
Distance between adjacent floater edge	1 m
Truss diameter	0.06 m
Truss number in one side	6

Table 3. Parameters of one floater.

Meaning	Value
Diameter of floater	3 m
Mass of floater	3000 kg
Draft of floater	1 m
Mass of floating arm	300 kg
Length of floating arm	7 m
Displacement	7.07 m³
Floater moment of inertia I_{xx}	2687 kg·m²
Floater moment of inertia I_{yy}	2687 kg·m²
Floater moment of inertia I_{zz}	3375 kg·m²

Table 4. Location of damping boxes (unit: m).

Damping Box 1	Damping Box 2	Damping Box 3	Damping Box 4	Damping Box 5
(8.35, 8.35)	(8.35, −8.35)	(−8.35, −8.35)	(−8.35, 8.35)	(0.0, 0.0)

4.1. The Influence of Floater Number on the Motion Response

In order to study the influence of the floater number on the power generation efficiency of the WEC platform, the motion responses of the floaters in the heave motion when the number of single-sided floaters from 1 to 4 are carried out through numerical simulation, and the calculation diagram is shown in Figure 6. The left floaters use $f_{li}(i = 1, 2, 3, 4)$ to represent the i-th floater on the left side of the platform, and the right floater uses $f_{ri}(i = 1, 2, 3, 4)$ to represent the i-th floater on the right side of the platform. Wave incident angles of 180 degrees and 90 degrees are used here to simulate the wave propagating along the floater arrangement direction and perpendicular to the arrangement direction, respectively.

Firstly, the heave motion responses of the floater are analyzed and compared under regular waves with different periods when a floater is arranged on one side, as shown in Figure 7. The response amplitude operator (RAO) in the y-axis indicates the motion characteristics of the floater in heave motion. The results indicate that when the wave propagates along the direction of the floater arrangement (wave incidence angle of 180 degrees), the motion response of the up-wave floater is greater than that of the back-wave floater, especially in the case of short-period wave action, because the damping boxes of the platform have a certain degree of reflection on the wave. When the wave period is 4 s, the wave force in the heave direction is small, so the heave motion in the period of 4 s is small. The natural period of the single floater is near 5.0 s, so the heave motion in the period of 5 s increases obviously due to the resonance. However, in the case of long-period wave action, the influence of the truss-type platform on the wave propagation becomes smaller, and the motion of the up-wave and back-wave floaters tends to be the same. When the wave

propagates perpendicular to the float arrangement direction (the wave incidence angle is 90 degrees), the motion amplitudes of the floaters on both sides are the same.

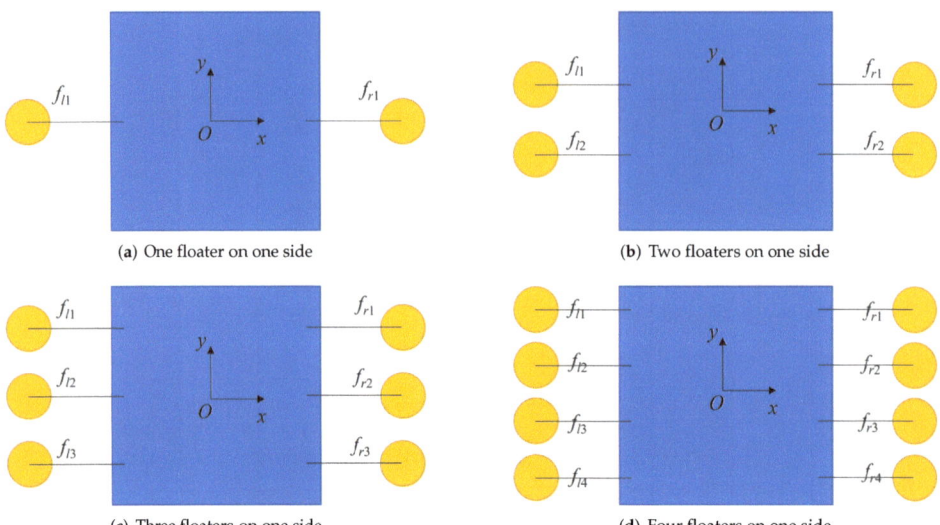

Figure 6. Sketch of floaters' arrangement for different cases.

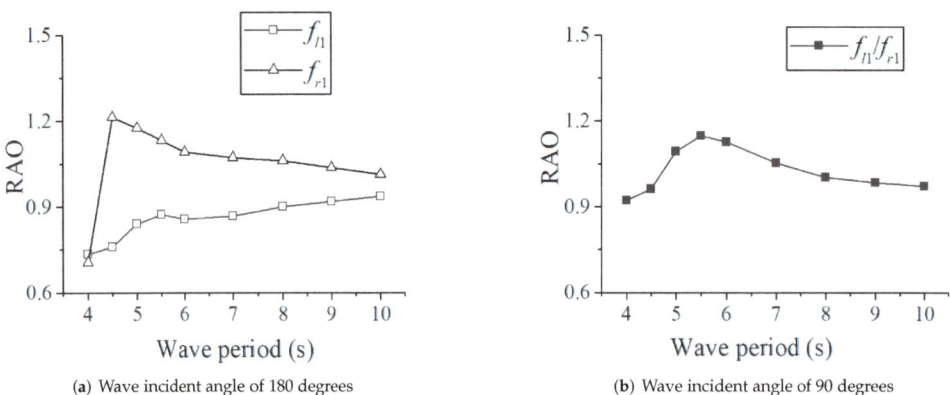

Figure 7. Heave motion response of floaters in different wave periods (one floater on one side).

Then, the results were analyzed when two floaters are arranged on one side, as shown in Figure 8. When the wave propagates along the floater arrangement direction, the heave motion results are similar to those of a single floater. When the wave propagates perpendicular to the floater arrangement direction, the front float has a certain degree of shielding effect on the rear floater, and the motion response is slightly greater than that of the rear floater; however, the overall difference is not large.

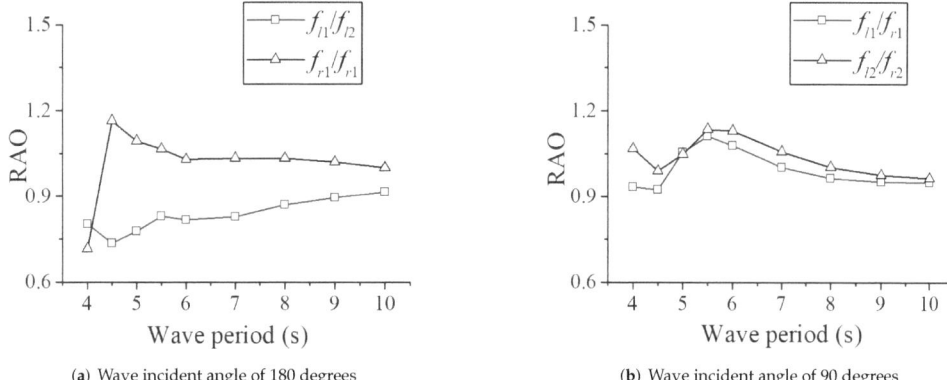

Figure 8. Heave motion response of floaters in different wave periods (two floaters on one side).

Further, the calculation results of three floaters and four floaters on one side were analyzed, as shown in Figures 9 and 10. The conclusion is similar to the previous one. When the wave propagates along the floater arrangement direction, the heave motion of the floaters on both sides is slightly larger than that of the floater in the middle, and the overall motion of the up-wave floater is slightly larger, while the motion of the back-wave floater is slightly smaller. The motion response tends to be consistent with an increase in wave period. When the wave propagates perpendicular to the floater arrangement direction, the motion trends of each float are similar, and the motion response of the front floater is slightly larger than that of the rear floater; however, the overall difference is not significant. Combined with the RAO comparison of different wave periods, when the wave period is small, the damping boxes and floaters influence the wave field, and the motion responses of the up-wave floater are larger than those of the back-wave floater. With an increase in wave period, the relative length of the structure and wavelength becomes small, so the motion trends of the up- and back-wave floaters tend to be consistent.

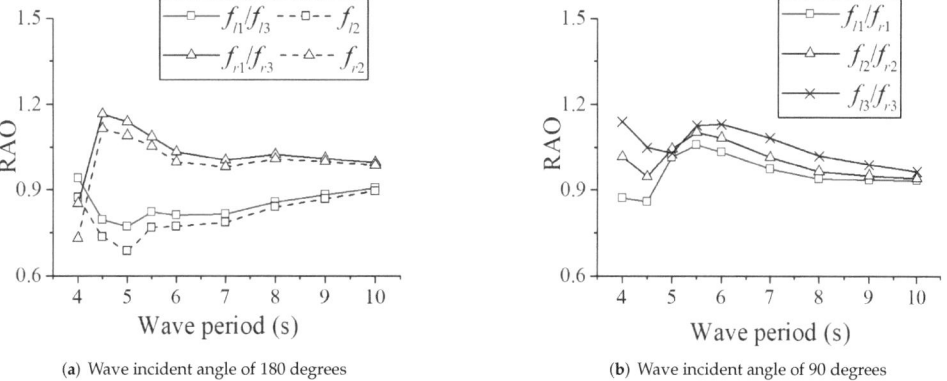

Figure 9. Heave motion response of floaters in different wave periods (three floaters on one side).

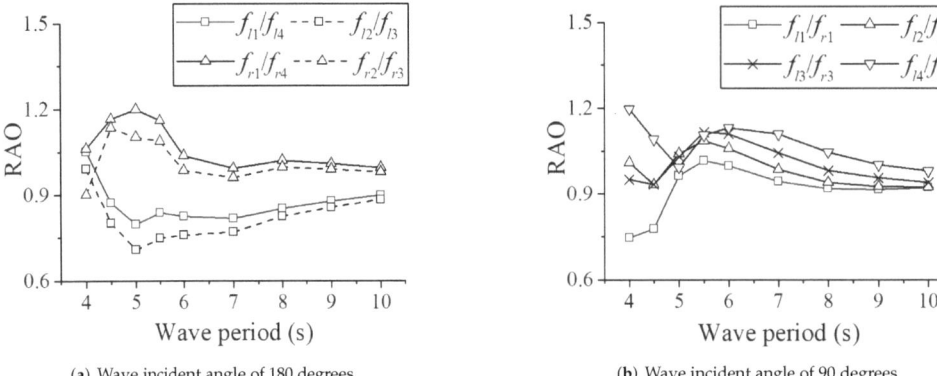

Figure 10. Heave motion response of floaters in different wave periods (four floaters on one side).

On the whole, since the platform adopts a truss-type structure; its existence has little effect on the wave field; the difference in the motion response of the floaters under the wave action in all directions is not particularly large; each floater has a good wave-following property. However, the size of the floater is relatively large, and the phenomenon of diffraction and scattering will occur when waves pass through the floater. The heave motion responses of the up-wave floater are larger than those of the back-wave floaters.

4.2. The Selection of Floater Arrangement

Based on the above four situations, the optimal floater arrangement is selected by considering the power generation efficiency and economy of the device as a whole. Under the conditions of 0° and 90° incident waves, the average RAOs of the floater are shown in Figure 11. The total average RAOs of the two directions are shown in Figure 11c. Whether the wave propagates along the floater arrangement direction or the wave propagates perpendicular to the floater arrangement direction, the RAOs of the floater are the largest in the case of a single floater arrangement in most cases, and with an increase in the number of arranged floaters, the average RAOs of the floater become smaller. When the wave period is 5 s, the heave motion responses of the floaters reach a maximum. This shows that within a certain range, although the existence of the floater has an influence on the motion response, the overall quality of the floater is small and the size of the float is not particularly large, so the effect is limited. The floaters as a whole show a very good wave-following property. Therefore, arranging as many floaters as possible can effectively improve the power efficiency of the WEC platform.

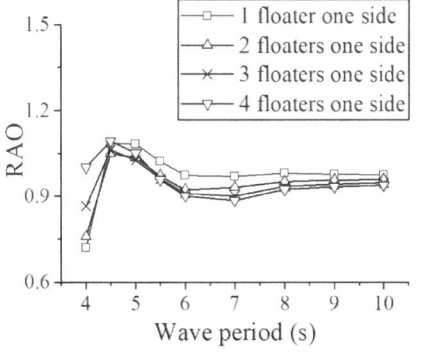
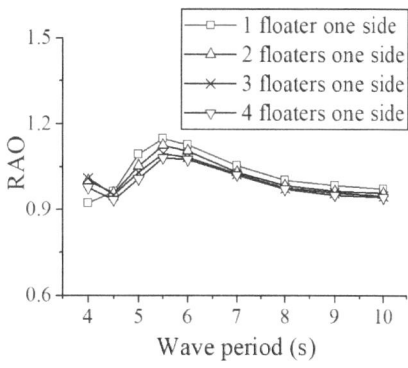

(a) Wave incident angle of 180 degrees (b) Wave incident angle of 90 degrees

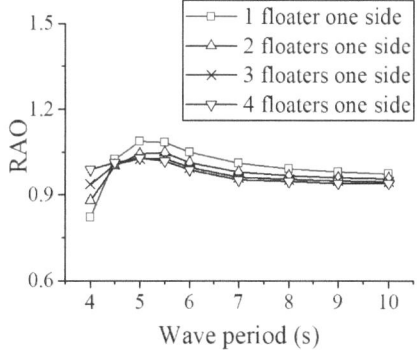

(c) Total average values of the two directions

Figure 11. Average heave motion response of floaters in different wave periods.

5. Calculation of Energy Utilization Efficiency of WEC Platform

5.1. Theory of Capture Efficiency

Wave potential energy and kinetic energy together make up the total wave energy stored per unit area in the length of a single wide wave crest line per unit wavelength. In Airy wave theory, it can be expressed as

$$E_p = E_k = \frac{1}{16}\rho g H^2 \lambda \tag{17}$$

where E_p denotes wave potential energy, E_k denotes wave kinetic energy, H denotes the wave height, and λ denotes wavelength. The average total wave energy of the peak line length of a single width within the unit wavelength is

$$E = \frac{E_p + E_k}{\lambda} = \frac{1}{8}\rho g H^2 \tag{18}$$

Therefore, the average total energy of waves in the rectangular area corresponding to the floater within a wave period is

$$E_f = E \cdot A_t = \frac{1}{8}\rho g H^2 A_t \tag{19}$$

where A_t denotes the rectangular area corresponding to the floater, which is the width of the float multiplied by the wavelength. The energy captured by the floater from the

ocean waves is mainly transformed into two parts, namely the kinetic energy and potential energy of the floater. The kinetic energy absorbed by the floater can be expressed as:

$$E_{fk}(t) = \frac{1}{2}(m+m_a)v^2 = \frac{1}{2}(m+m_a)\omega^2 A^2 \sin^2(\omega t + \varphi) \tag{20}$$

The potential energy absorbed by the floater can be expressed as:

$$E_{fp}(t) = \frac{1}{2}kx^2 = \frac{1}{2}\rho g A_w x^2 = \frac{1}{2}\rho g A_w A^2 \cos^2(\omega t + \varphi) \tag{21}$$

Therefore, the total wave energy absorbed by the floater system can be expressed as:

$$E_{fz}(t) = E_{fk}(t) + E_{fp}(t) = \frac{1}{2}(m+m_a)\omega^2 A^2 \sin^2(\omega t + \varphi) + \frac{1}{2}\rho g A_w A^2 \cos^2(\omega t + \varphi) \tag{22}$$

where m denotes the floater mass, m_a denotes the added mass of the floater, and A_w denotes the water line area of the floater. The wave energy captured by the oscillating floater system within a wave period is:

$$E_{f2} = \int_0^{WT} E_{fz}(t) dt \tag{23}$$

Therefore, the energy capture efficiency of the WEC platform can be expressed as the ratio of the total energy absorbed by the floater to the total energy of the wave in the width area of the floater, that is:

$$\eta_f = \frac{E_{f2}}{E_f} \tag{24}$$

5.2. Capture Efficiency of WEC Platform

In the multi-floater WEC platform system, the heave motion of the floater relative to the platform overcomes the damping force to produce useful power, and the power take-off (PTO) system converts the mechanical energy of floater movement into electrical energy. In fact, in order to simplify the model, we use the linear damping model to analyze the damping force of the PTO system. In addition, there are some studies that have used the same linear damping model to consider the PTO system [27,28]. We simulate the damping force in the PTO system, fd, using the following equation

$$f_d = -C_{d,f} \cdot v \tag{25}$$

where v is the heaving velocity of the floater and $C_{d,f}$ is a linear damping coefficient, which represents the performance of the PTO system. The absorbed power is represented by

$$P = \frac{1}{WT}\int_0^{WT} f_d \cdot v dt = \frac{1}{WT}\int_0^{WT} -C_{d,f} \cdot v^2 dt \tag{26}$$

where f_d is the damping force and WT is the wave period. The different damping coefficients $C_{d,f}$ were selected by testing several damping coefficients, before the best value, which yields the maximum absorption power, was eventually determined.

In this WEC platform, the floater is connected to the platform through the floating arm, which makes the floater have a heave movement in the wave field. The end of the floating arm pushes the PTO power output device through the lever principle to overcome the PTO damping. The work in this part is the energy used by the power generation system of the device. In the process of capturing wave energy by the power generation system, the influence of PTO damping on energy capture efficiency is very significant. The value of PTO damping can change the heave motion response and motion velocity of the floater, thus affecting the capture of wave energy by the floater. Therefore, it is particularly necessary to study the impact of different linear PTO damping on the capture efficiency

of the device. In the numerical simulation, a constant "C" is introduced as the damping coefficient to adjust the damping force. Through the simulation of the multi-floater WEC platform with different linear damping coefficients C, the influence of different linear PTO damping coefficients on capture efficiency is explored. The linear damping coefficient C is introduced in the equation and the total absorbed power of the floater in the case of damping coefficient C.

In the numerical simulation, the linear damping coefficient in the heave direction is added to one floater. The wave propagates along the floater arrangement direction, and the wave amplitude is 0.5 m; the wave period is 5 s when the floater average RAO is the maximum. The velocities of the floater in the heave motion with time are calculated here. The velocity time histories under some different linear damping coefficients C are shown in Figures 12–15. The results show that the velocity of the floater decreases significantly with an increase in the linear damping coefficient.

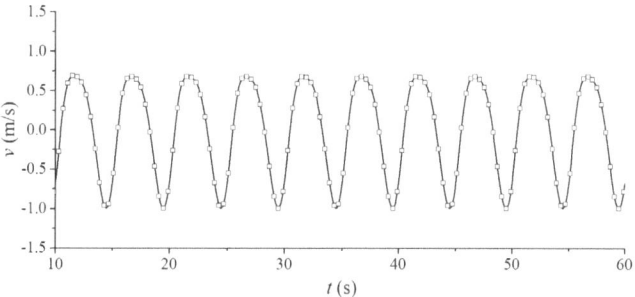

Figure 12. Velocity time histories of the floater when linear damping coefficient C is 3000 N/(m/s).

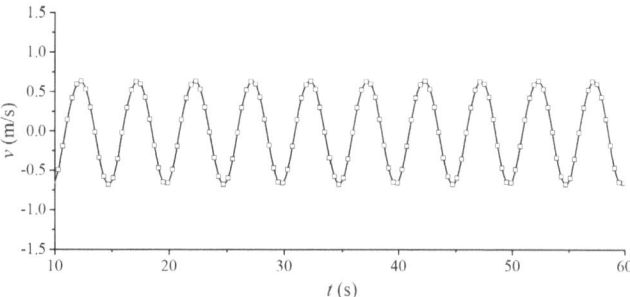

Figure 13. Velocity time histories of the floater when linear damping coefficient C is 20,000 N/(m/s).

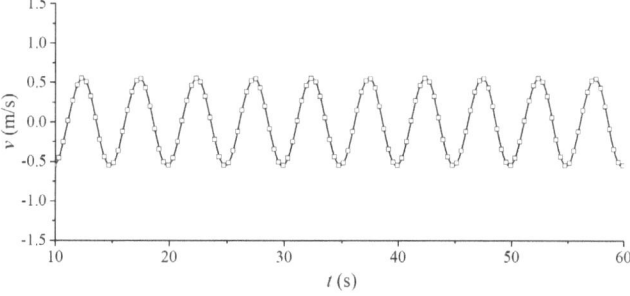

Figure 14. Velocity time histories of the floater when linear damping coefficient C is 30,000 N/(m/s).

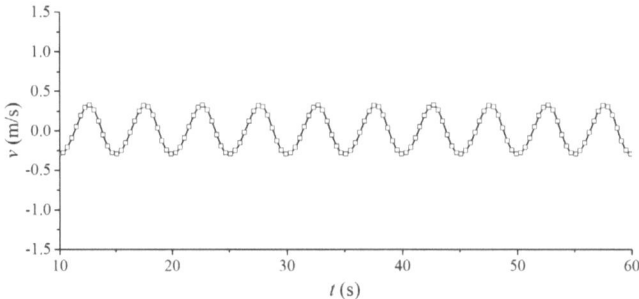

Figure 15. Velocity time histories of the floater when linear damping coefficient C is 70,000 N/(m/s).

Furthermore, the power of the floater to capture wave energy under different linear PTO damping coefficients can be calculated, as shown in Figure 16a. Therefore, the relationship between the energy capture efficiency of the floater and the linear PTO damping coefficient C is calculated as shown in Figure 16b. With an increase in the linear damping coefficient C, the capture efficiency of the floater first increases, then decreases, and reaches the maximum when C = 30,000 N/(m/s); thus, it is concluded that C = 30,000 N/(m/s) is the optimum PTO damping coefficient for the WEC platform. The wave capture efficiency of each floater of the WEC platform is further obtained, as shown in Table 5.

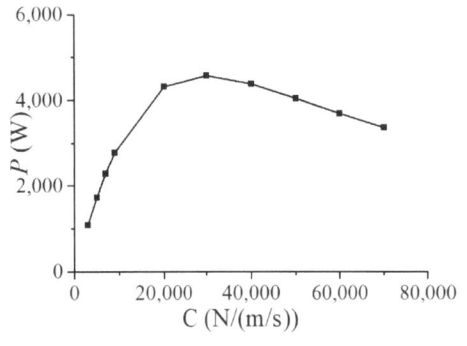
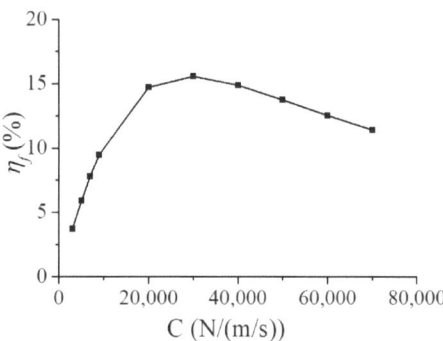

(a) Generation power of the floater (b) Capture efficiency of the floater

Figure 16. Generation power and capture efficiency of the floater to capture wave energy under different linear PTO damping coefficients.

Table 5. Capture efficiency of each floater.

Floater number $\eta_f(\%)$	f_{l1} 5.38	f_{l2} 4.07	f_{l3} 4.07	f_{l4} 5.38
Floater number $\eta_f(\%)$	f_{r1} 15.55	f_{r2} 13.80	f_{r3} 13.80	f_{r4} 15.55

6. Conclusions

Based on potential flow theory, the mathematical model of the interaction between wave and WEC platform was established by using the high-order boundary element method. The motion response of each floater of the multi-floater truss-type WEC platform was studied. The influence of the number of floaters and the arrangement of floaters on the motion of the floater was analyzed. The effect of the hydraulic cylinder on the float was simulated by linear damping to generate power. The multi-float truss-type WEC generation device was simulated, and the following conclusions were obtained:

1. Since the platform is composed of a truss structure, the floaters' motion responses under wave action in all directions are very close, and each floater has good wave-following performance. The motion response of the up-wave floater is slightly greater than that of the back-wave floater, and the motion response of the side floaters is slightly greater than that of the middle floater.
2. With an increase in the number of floaters arranged at one side, the average RAO of the floats is smaller, but the overall difference is small. Multiple floaters as a whole show very good wave-following performance. Therefore, if conditions permit, arranging as many floats as possible can effectively improve the power generation efficiency of the platform.
3. The power and efficiency of a single float first increase and then decrease with the increase of linear damping under the regular wave action. When the damping is 30,000 N/(m/s), the power generation efficiency is the highest, and the capture efficiency of the whole platform is 9.7%

Author Contributions: Conceptualization, R.J. and J.W.; methodology, R.J.; software, J.W.; validation, J.W.; writing—original draft preparation, R.J.; writing—review and editing, B.G. and Z.L.; funding acquisition, H.C. All authors have read and agreed to the published version of the manuscript.

Funding: This work was supported by the Application Demonstration of High-Reliability Marine Energy Supply Equipment (GHME2018SF02); Basic Funding of the Central Public Research Institutes (TKS20200406, TKS20210108); Science and Technology Research and Development Project of CSCES (CSCEC-2020-Z-21).

Institutional Review Board Statement: Not applicable.

Informed Consent Statement: Informed consent was obtained from all subjects involved in the study.

Conflicts of Interest: The authors declare no conflict of interest.

References

1. Zhang, D.H.; Li, W.; Lin, Y.G. Wave energy in China: Current status and perspectives. *Renew. Energy* **2009**, *34*, 2089–2092. [CrossRef]
2. Nielsen, K.; Smed, P.F. Point absorber-optimization and survival testing. In Proceedings of the Third European Wave Energy Conference, Patras, Greece, 30 September–2 October 1998; Volume 18.
3. Prado, M. Archimedes wave swing (AWS). In *Ocean Wave Energy*; Springer: Berlin, Germany, 2008; pp. 297–304.
4. Cleason, L.; Forsberg, J.; Ryler, A.; Sjöström, B.O. Contribution to the theory and experience of energy production and transmission from the buoy-concept. In Proceedings of the 2nd International Symposium on Wave Energy Utilization, Trondheim, Norway, 22–24 June 1982; pp. 345–370.
5. Chang-Lei, M.A.; Xia, D.W.; Wang, M. Review on the Progress of Marine Renewable Energy Technologies in the World. *J. Ocean Technol.* **2017**, *36*, 70–75. (In Chinese)
6. Korde, U.A. On controlled oscillating bodies with submerged reaction mass. In Proceedings of the OMAE 2003 22nd International Conference on Offshore Mechanics and Arctic Engineering, Cancun, Mexico, 8–13 June 2003.
7. Caska, A.J.; Finnigan, T.D. Hydrodynamic characteristics of a cylindrical bottom-pivoted wave energy absorber. *Ocean Eng.* **2008**, *35*, 6–16. [CrossRef]
8. Zhao, X.L.; Ning, D.Z.; Zhang, C.W.; Kang, H.G. Hydrodynamic Investigation of an Oscillating Buoy Wave Energy Converter Integrated into a Pile-Restrained Floating Breakwater. *Energies* **2017**, *10*, 712. [CrossRef]
9. Cao, C. Investigation on Hydrodynamic Performance of Wave Energy Float Pendulum System. Master's Thesis, Suzhou University, Suzhou, China, 2019. (In Chinese)
10. Zhou, Z. Mooring Hydrodynamics Analysis of Oscillating Float Wave Power Generator. Master's Thesis, Ningbo University, Ningbo, China, 2019. (In Chinese)
11. Zhou, B.H. The Research of Pitching Float Type Wave Energy Conversion Device. Master's Thesis, Harbin Engineering University, Harbin, China, 2018.
12. Zhang, H.M.; Zhou, B.Z.; Vogel, C.; Willden, R.; Zang, J.; Geng, J. Hydrodynamic performance of a dual-floater hybrid system combining a floating breakwater and an oscillating-buoy type wave energy converter. *Appl. Energy* **2020**, *259*, 114212. [CrossRef]
13. Finnegan, W.; Rosa-Santos, P.; Taveira-Pinto, F.; Geooins, J. Development of a numerical model of the CECO wave converter using computational fluid dynamics. *Ocean Eng.* **2021**, *219*, 108416. [CrossRef]
14. Luan, Z.X.; He, G.H.; Zhang, Z.G.; Jing, P.L.; Jin, R.J.; Geng, B.L.; Liu, C.G. Study on the optimal wave energy absorption power of a float in waves. *J. Mar. Sci. Eng.* **2019**, *7*, 269 [CrossRef]

15. Liu, Q.L. Study on the Wave Energy Conversion Characteristics of an Array of Floating Point Absorbers. Ph.D. Thesis, Tsinghua University, Beijing, China, 2016. (In Chinese)
16. Yang, S.H.; He, G.Y.; He, H.Z. Response and Efficiency Analysis of Oscillating Float Array Based on Hydrodynamic Simulation. In Proceedings of the State Oceanic Administration of the State Oceanic Technology Center Proceedings of the 4th China Marine renewable Energy Development Annual Conference and Forum, Suzhou, China, 21–23 October 2015; pp. 58–64. (In Chinese)
17. Chrasekaran, S.; Sricharan, V.V.S. Numerical study of bean-float wave energy converter with float number parametrization using WEC-Sim in regular waves with the Levelized Cost of Electricity assessment for Indian sea states. *Ocean Eng.* **2021**, *237*, 109591. [CrossRef]
18. Marchesi, E.; Negri, M.; Malavasi, S. Development and analysis of a numerical model for a two-oscillating-body wave energy converter in shallow water. *Ocean Eng.* **2020**, *214*, 107765. [CrossRef]
19. He, G.H.; Luan, Z.X.; Jin, R.J.; Zhang, W.; Wang, W.; Zhang, Z.G.; Jing, P.L.; Liu, P.F. Numerical and experimental study on absorber-type wave energy converters concentrically arranged on an octagonal platform. *Renew. Energy* **2022**, *188*, 504–523. [CrossRef]
20. Negri, M.; Malavasi, S. Wave Energy Harnessing in Shallow Water through Oscillating Bodies. *Energies* **2018**, *11*, 2730. [CrossRef]
21. Ramadanad, A.; Mohamedbc, M.H.; Gabbard, H.A. Experimental analysis of an enhanced design of float with inverted cup for wave energy conversion. *Ocean Eng.* **2022**, *249*, 110910. [CrossRef]
22. Moreno, E.C.; Stansby, P.K. The 6-float wave energy converter M4: Ocean basin tests giving capture width, response and energy yield for several sites. *Renew. Sustain. Energy Rev.* **2019**, *104*, 307–318. [CrossRef]
23. Stansby, P.K.; Moreno, E.C. Hydrodynamics of the multi-float wave energy converter M4 with slack moorings: Time domain linear diffraction-radiation modeling with mean force and experimental comparison. *Appl. Ocean Res.* **2020**, *97*, 102070. [CrossRef]
24. Santo, H.; Taylor, P.H.; Stansby, P.K. The performance of the three-float M4 wave energy converter off Albany, on the south coast of western Australia, compared to Orkney (EMEC) in the U.K. *Renew. Energy* **2020**, *146*, 444–459. [CrossRef]
25. Jin, R.J.; Gou, Y. Motion response analysis of large-scale structures with small-scale cylinders under wave action. *Ocean. Eng.* **2018**, *115*, 65–74. [CrossRef]
26. Choi, Y.; Hong, S. An Analysis of Hydrodynamic Interaction of Floating Multi-Body Using Higher-Order Boundary Element Method. In Proceedings of the International Offshore and Polar Engineering Conference, Kitakyushu, Japan, 26–31 May 2002; Volume 12, pp. 303–308.
27. Chrasekaran, S.; Sricharan, V.V.S. Numerical analysis of a new multi-body floating wave energy converter with a linear power take-off system. *Renew. Energy* **2020**, *159*, 250271.
28. Kim, K.H.; Park, S.; Kim, J.R.; Cho, I.H.; Hong, K. Numerical and experimental analyses on motion responses on heaving point Absorbers connected to large semi-submersibles. *Processes* **2021**, *9*, 1363. [CrossRef]

Article

Research on Size Optimization of Wave Energy Converters Based on a Floating Wind-Wave Combined Power Generation Platform

Xianxiong Zhang [1], Bin Li [1], Zhenwei Hu [1], Jiang Deng [2], Panpan Xiao [2] and Mingsheng Chen [2,3,*]

[1] Department of Engineering, Poly Changda Engineering Co., Ltd., Guangzhou 510620, China
[2] School of Naval Architecture, Ocean and Energy Power Engineering, Wuhan University of Technology, Wuhan 430063, China
[3] Sanya Science and Education Innovation Park of Wuhan University of Technology, Sanya 572025, China
* Correspondence: mschen@whut.edu.cn

Abstract: Wind energy and wave energy often co-exist in offshore waters, which have the potential and development advantages of combined utilization. Therefore, the combined utilization of wind and waves has become a research hotspot in the field of marine renewable energy. Against this background, this study analyses a novel integrated wind-wave power generation platform combining a semi-submersible floating wind turbine foundation and a point absorber wave energy converter (WEC), with emphasis on the size optimization of the WEC. Based on the engineering toolset software ANSYS-AQWA, numerical simulation is carried out to study the influence of different point absorber sizes on the hydrodynamic characteristics and wave energy conversion efficiency of the integrated power generation platform. The well-proven CFD software STAR CCM+ is used to modify the heaving viscosity damping of the point absorber to study the influence of fluid viscosity on the response of the point absorber. Based on this, the multi-body coupled time-domain model of the integrated power generation platform is established, and the performance of the integrated power generation platform is evaluated from two aspects, including the motion characteristics and wave energy conversion efficiency, which provides an important reference for the design and optimization of the floating wind-wave power generation platform.

Keywords: floating wind-wave power generation platform; WEC; numerical simulation; wave power conversion efficiency; viscous heaving damping correction

Citation: Zhang, X.; Li, B.; Hu, Z.; Deng, J.; Xiao, P.; Chen, M. Research on Size Optimization of Wave Energy Converters Based on a Floating Wind-Wave Combined Power Generation Platform. *Energies* **2022**, *15*, 8681. https://doi.org/10.3390/en15228681

Academic Editor: Guanghua He

Received: 20 October 2022
Accepted: 16 November 2022
Published: 18 November 2022

Publisher's Note: MDPI stays neutral with regard to jurisdictional claims in published maps and institutional affiliations.

Copyright: © 2022 by the authors. Licensee MDPI, Basel, Switzerland. This article is an open access article distributed under the terms and conditions of the Creative Commons Attribution (CC BY) license (https://creativecommons.org/licenses/by/4.0/).

1. Introduction

In recent years, the integration of floating wind turbines and Wave Energy Converters (WECs) onto a single platform for combined utilization has become a research hotspot. Some European countries first launched a series of ocean energy projects, such as Poseidon P37, W2Power and WindWavefloat, aiming at promoting the development and application of the combined utilization platform of wind and wave energies and exploring the various combination forms of WECs integrated with offshore wind turbines. Peiffer et al. [1] proposed a hybrid integrated platform consisting of the WindFloat three-column semi-submersible floating platform and a point absorber WEC. Additionally, Aubault et al. [2] proposed a concept that integrates the Oscillating Water Column (OWC) WECs with the WindFloat floating wind turbine. Their studies show that by adding WEC to WindFloat, the mooring and power infrastructure can be shared, and the overall economic cost can be effectively reduced. Soulard and Babarit [3] proposed a hybrid platform that combines a 5 MW wind turbine with floating Oscillating Surge Wave Energy Converters (OSWECs) and preliminarily evaluated its response and power generation using both the frequency-domain and time domain approaches. Their studies prove that the power generation time–history curve of the combined platform is smoother than that of the single platform.

Muliawan et al. [4] and Luan et al. [5] proposed the Spar-Torus Combination (STC) concept and the Semi-submersible Flap Combination (SFC) concept, respectively. Their research shows that the integration of WECs onto the floating wind turbine foundation can improve the overall power generation and reduce the dynamics of the wind turbine foundation to a certain extent. Gaspar et al. [6] investigated the concept of using WECs to compensate for the dynamics of a hybrid wind-wave power generation platform. Their research shows that the ballast water system of the platform assisted by the WECs can expand the working sea-state range to a certain extent.

The aforementioned studies show that the WEC can cooperate well with the floating wind turbine. Further, in the aspect of WEC design, based on the SFC concept [5], Michailides et al. [7] further investigated the influence of the number of flap-type WECs on the stability, motion, and internal load of the combined power generation platform. The results show that the combined operation of the rotating flaps can increase the overall power generation without greatly affecting the response of the semi-submersible wind turbine foundation. For the STC concept, Wan et al. [8] and Muliawan et al. [9] further studied the responses of the STC platform under extreme sea-states, which provided an important reference for the design of WEC to be integrated with a Spar-type floating wind turbine. Hallak et al. [10] studied the linear hydrodynamic interaction of a hybrid power generation platform combining DeepCWind semi-submersible platform with conical point absorbers. Their results show that the mechanical coupling of the combined platform can amplify the heave response of the platform. Thus, a more realistic physical model should be considered in the power take-off (PTO) system design, such as the selection of the stiffness and damping of the PTO system. Wang et al. [11] proposed a marine energy structure composed of a 5 MW bracket-free semi-submersible floating wind turbine and a pendulum WEC. They determined its reasonable draft, size, and PTO damping coefficient by numerical simulations and also studied the influence of viscosity on the dynamics of the coupled system. González et al. [12] performed parametric studies to investigate the rigid body dynamics of a hybrid platform consisting of a semi-submersible foundation and a point absorber WEC. Their research shows that the relatively slender point absorber and larger heave plate of the semi-submersible foundation can effectively reduce the dynamics of the hybrid system. Sun et al. [13] carried out basin tests to evaluate the concept of combining a semi-submersible floating wind turbine platform with six cone-bottom point absorber WECs. Their results show that the point absorbers should be staggered, and the energy conversion efficiency of multiple point absorbers is higher than that of a single point absorber in the low wave period range. Hu et al. [14] applied the frequency-domain hydrodynamic model with viscosity correction to optimize the size and layout of the WECs in a combined wind-wave power generation platform. Lee et al. [15] investigated the coupled dynamics of the floating platform with multiple WECs by considering multi-body hydrodynamic interactions. Hantoro et al. [16] applied both numerical and experimental methods to investigate the relation between the response of the trimaran-type pontoon array WEC and its pendulum system. This study shows that the differences in array arrangement and wave period affect the pitching motions of the pontoon. Similar to the concept proposed by Peiffer et al. [1], Chen et al. [17] proposed an integrated wind-wave power generation platform with a point absorber WEC placed inside the semi-submersible wind turbine foundation. This concept aims to benefit from the near-trapping effects caused by the multiple columns of the semi-submersible foundation to enhance the wave power generation of the point absorber WEC, which uses the relative heaving motions between the foundation and WEC to generate electricity with the PTO system placed on the foundation. Following this concept, Chen et al. [18] established a dynamic coupling method between AQWA and Fast to perform a fully coupled analysis of the integrated wind-wave power generation platform, which confirmed the feasibility and advantages of this concept. However, none of the above studies has systematically analyzed and optimized the size of WEC to be integrated with a floating wind turbine. Based on the concept proposed by Chen et al. [17], this study aims to optimize the size of the WEC based on frequency-domain hydrodynamic

analyses. In this study, the well-proven hydrodynamic software ANSYS-AQWA is used to carry out frequency-domain hydrodynamic analyses of the multi-body system. The influence of the size of the point absorber WEC on the hydrodynamic characteristics and wave energy conversion efficiency of the integrated wind-wave power generation platform is studied. On this basis, CFD software STAR CCM+ is used to correct the heaving damping viscosity of the point absorber WEC. In addition, a multi-body coupled time-domain model, considering only the heaving motions of the integrated platform, is also established in ANSYS-AQWA. The responses and wave power generation efficiency of the integrated platform are explored in both regular and irregular waves. Compared with a single WEC, it may be concluded that the dynamic and hydrodynamic coupling effects of the integrated platform can improve the WEC's power generation efficiency within a certain wave frequency range, and it makes up the research gap of Chen et al. [17,18] in the optimization of WEC size. Besides, this study can provide an important reference for the design and optimization of a floating wind-wave power generation platform.

2. Mathematical Model

The analyzed integrated platform is based on an OC4 semi-submersible floating wind turbine foundation, as reported by Robertson et al. [19] and Cheng et al. [20]. The particulars of the semi-submersible foundation are summarized in Table 1 [19,20]. Based on the concept by Chen et al. [17,18], a point absorber WEC is placed in the center of the semi-submersible platform, and the PTO system is placed directly above the point absorber, as shown in Figure 1. Figure 1a shows the hydrodynamic model of the integrated platform, in which the connecting members of the semi-submersible foundation that have little effect on the hydrodynamic interaction are omitted to simplify the frequency-domain hydrodynamic analysis. The PTO system is demonstrated in Figure 1b. Under the wave action, the point absorber and the semi-submersible platform would move vertically, and the mechanical energy of this relative vertical motion is converted into electrical energy by the PTO system. In the following discussions, for simplicity, the point absorber WEC and the semi-submersible foundation are designated as "floater" and "platform", respectively.

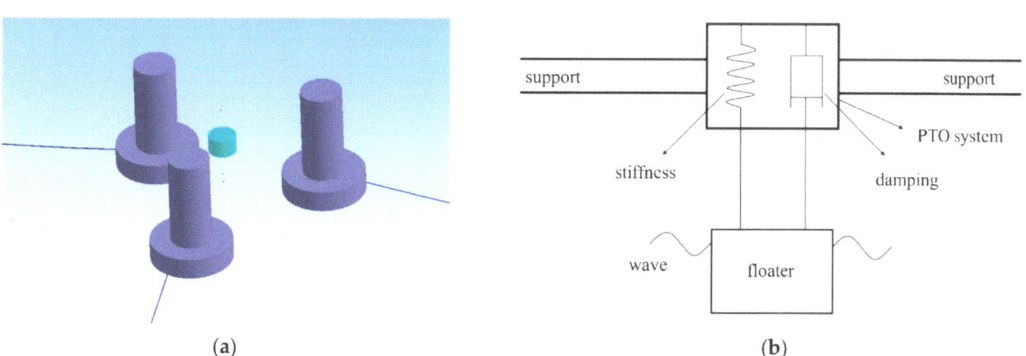

Figure 1. Schematic diagram of floating wind-wave combined power generation platform: (a) hydrodynamic model in ANSYS-AQWA; (b) simplified model of PTO system [17,18].

Table 1. Particulars of OC4 semi-submersible floating wind turbine foundation [19,20].

Geometric Parameter	Value
Draft of platform SWL (m)	20
Elevation of upper columns above SWL (m)	12
Spacing between offset columns (m)	50
Length of upper columns (m)	26

Table 1. Cont.

Geometric Parameter	Value
Length of base columns (m)	6
Diameter of base column (m)	6.5
Diameter of upper columns (m)	12

2.1. Frequency-Domain Model

Considering the coupling effect of floater (point absorber) and platform (semi-submersible foundation) through a linear PTO system and only the heaving motion of the hybrid system, the hydrodynamic coupling model in the frequency-domain has the following expression:

$$\left\{-\omega^2\begin{bmatrix}M_{33}+A_{33}(\omega) & A_{39}(\omega)\\A_{93}(\omega) & M_{99}+A_{99}(\omega)\end{bmatrix}-i\omega\begin{bmatrix}B_{33}(\omega)+B_{pto}+B_v & B_{39}(\omega)-B_{pto}\\B_{93}(\omega)-B_{pto} & B_{99}(\omega)+B_{pto}\end{bmatrix}\right.\\\left.+\begin{bmatrix}C_{33}+K_{pto} & -K_{pto}\\-K_{pto} & C_{99}+K_{pto}\end{bmatrix}\right\}\begin{bmatrix}\widehat{x}_3(i\omega)\\\widehat{x}_9(i\omega)\end{bmatrix}=\begin{bmatrix}\widehat{f}_3^{exc}(i\omega)\\\widehat{f}_9^{exc}(i\omega)\end{bmatrix} \quad (1)$$

where, M_{33}, $A_{33}(\omega)$, $B_{33}(\omega)$, C_{33}, $\widehat{x}_3(i\omega)$ and $\widehat{f}_3^{exc}(i\omega)$ are rigid mass matrix, added mass, radiation damping, hydrostatic restoring force stiffness matrix, response amplitude operator (RAO) and wave exciting force of the floater in the heave direction, respectively; M_{99}, $A_{99}(\omega)$, $B_{99}(\omega)$, C_{99}, $\widehat{x}_9(i\omega)$ and $\widehat{f}_9^{exc}(i\omega)$ are rigid mass matrix, added mass, radiation damping, hydrostatic restoring force stiffness matrix, RAO and wave exciting force of the platform in the heave direction; $A_{39}(\omega)$ and $A_{93}(\omega)$ are the coupling term of added mass between the floater and platform; $B_{39}(\omega)$ and $B_{93}(\omega)$ are the coupling term of radiation damping between the floater and platform; B_{pto}, K_{pto} and B_v are the PTO damping, PTO stiffness and viscous damping of the floater in heave mode of motion, respectively.

The viscous damping of the floater in heave model of motion can be generally obtained according to the free attenuation curve [18]. The attenuation coefficient can be calculated based on the following equation:

$$\kappa = \frac{\ln X_1 - \ln X_{N+1}}{2\pi N} \quad (2)$$

where, N is the number of motion attenuation and X_i is the amplitude of the ith motion.

After calculating the free attenuation coefficient, the total damping coefficient in the heave mode of motion can be obtained:

$$B_{vis} = 2\kappa\sqrt{C_{33}[M_{33}+A_{33}]} \quad (3)$$

Tom [21] proved in the experiment that the change in wave frequency has little influence on the viscous effect. Therefore, the total damping coefficient at the natural frequency can be calculated by Eq. (3). Combined with Tom's conclusion, the additional viscous damping can be approximated: [22]

$$B_v = B_{vis} - B_n \quad (4)$$

where, B_n is the radiation damping of the floater at the resonance frequency.

The frequency-generating power of WEC can be expressed as:

$$P_{ave} = \frac{1}{2}B_{pto}\omega^2\left|\widehat{x}_3(i\omega)-\widehat{x}_9(i\omega)\right|^2 \quad (5)$$

2.2. Time-Domain Model

The simplified schematic diagram of the floating wind-wave combined platform is shown in Figure 1. Based on Cummins equation which describes the time-domain motion of the floating body [23], a simplified time-domain model considering only heave motions and the PTO system of the integrated platform can be established in ANSYS-AQWA:

$$\begin{bmatrix} M_{33} + A_{33}(\infty) & A_{39}(\infty) \\ A_{93}(\infty) & M_{99} + A_{99}(\infty) \end{bmatrix} \begin{bmatrix} \ddot{x}_3 \\ \ddot{x}_9 \end{bmatrix} + \begin{bmatrix} \int_0^t K_{33}(t-\tau)\dot{x}_3(\tau)d\tau & \int_0^t K_{39}(t-\tau)\dot{x}_9(\tau)d\tau \\ \int_0^t K_{93}(t-\tau)\dot{x}_3(\tau)d\tau & \int_0^t K_{99}(t-\tau)\dot{x}_9(\tau)d\tau \end{bmatrix} + \begin{bmatrix} B_{pto} + B_v & -B_{pto} \\ -B_{pto} & B_{pto} \end{bmatrix} \begin{bmatrix} \dot{x}_3 \\ \dot{x}_9 \end{bmatrix} + \begin{bmatrix} C_{33} + K_{pto} & -K_{pto} \\ -K_{pto} & C_{99} + K_{pto} \end{bmatrix} \begin{bmatrix} x_3 \\ x_9 \end{bmatrix} = \begin{bmatrix} f_3(t) \\ f_9(t) \end{bmatrix} + \begin{bmatrix} 0 \\ f_m(t) \end{bmatrix} \quad (6)$$

where $A(\infty)$ is the approximate added damping coefficient; $K(t)$ is impulse response functions; x is the displacement of the floater in the heave direction; $f(t)$ and $f_m(t)$ are, respectively, the wave exciting force matrix and the mooring load matrix; Subscripts 3 and 9 refer to the degree of freedom in the heave of the floater and platform respectively.

Among them, the infinite frequency added mass and approximate damping coefficient of the floating body can be obtained by frequency-domain analysis software ANSYS-AQWA. And then, based on the conversion relationship between the hydrodynamic coefficient and impulse response function proposed by Ogilvie [24], the infinite frequency added mass coefficient is calculated according to Equations (7) and (8).

$$K(t) = \frac{2}{\pi} \int_0^\infty B(\omega) \cos(\omega t) d\omega \quad (7)$$

$$A(\infty) = A(\omega) + \frac{1}{\omega} \int_0^\infty K(t) \sin \omega t \, dt \quad (8)$$

According to the time–history curve of speed, the average generated power can be calculated as follows [25]:

$$P_{ave} = \frac{1}{t_p} \int_0^{t_p} B_{pto} \dot{x}_3^2(\tau) d\tau \quad (9)$$

where, t_p is the calculation duration of multiple exercise cycles.

3. Hydrodynamic Analysis in Frequency-Domain

3.1. Viscosity Correction for Hydrodynamic Calculation

To calculate the viscosity coefficient of the floater, CFD software STAR CCM+ is used to carry out the free decay test. According to the test results and based on the viscosity correction method [22], the viscosity coefficient of the floater falling freely from a certain height on the water surface is calculated. In the experiment, a simplified three-dimensional model of an oscillating float is adopted. Because of the symmetry of the model, only half of the model is used for the calculation to improve the calculation efficiency. In this study, it is necessary to simulate the real motion of the oscillating floater for a long time and a long distance, and based on the consideration of simulation accuracy, the physical model adopts the K-ε turbulence model and uses overlapping mesh technology to capture the motion trajectory of the floater. The volume of fluid (VOF) method is used to capture the liquid level change at the interface between water and air because the floater is in the two-phase flow of water and air. The calculation domain used in this study includes the background area of the simulated water area and the overlapping area of simulated floater motion, in which the background area is a cuboid with enough width and depth, and the overlapping area is obtained by subtracting the float itself from the smaller cylinder that wraps the float. The flow field information is transmitted between them by linear

interpolation. When setting boundary conditions in the region of the background, the top and bottom are taken as velocity inlets, the middle and longitudinal sections of the float are divided into symmetrical planes, and the other boundaries are all pressure outlets. During the simulation, the waves generated by the movement of the float will spread all around. To eliminate the influence of wave reflection on the results, damping to eliminate wave are set on the plane with the boundary condition of the pressure outlet. When meshing, the grids near the float and the free surface are properly encrypted, which makes the simulation results more accurate.

To verify the accuracy of the numerical simulation results, the free decay test is carried out on the cylindrical floater studied by Tom [21] and Chen [18,26]. The diameter of the cylindrical floater is 0.273 m, and the draft is 0.6123 m. The calculation domain and boundary conditions are set, as shown in Figure 2a. To eliminate the influence of cell size on simulation results, three sets of cells, coarse (1.43 million cells), medium (1.95 million cells) and fine (2.85 million cells), are selected for mesh convergence analysis, with the medium-sized cell set as shown in Figure 2b. The comparison of the simulation results of three sets of cells is shown in Figure 2c, and the result shows that the difference between the medium-sized cell and the fine-sized cell is very small, which indicates that the cell has converged. Therefore, on the premise of ensuring calculation accuracy and improving calculation efficiency, the medium-sized cell is used for the subsequent numerical simulation. Then, the CFD numerical simulation results are compared with the experimental data, as shown in Figure 2d. The result shows that the numerical simulation is accurate.

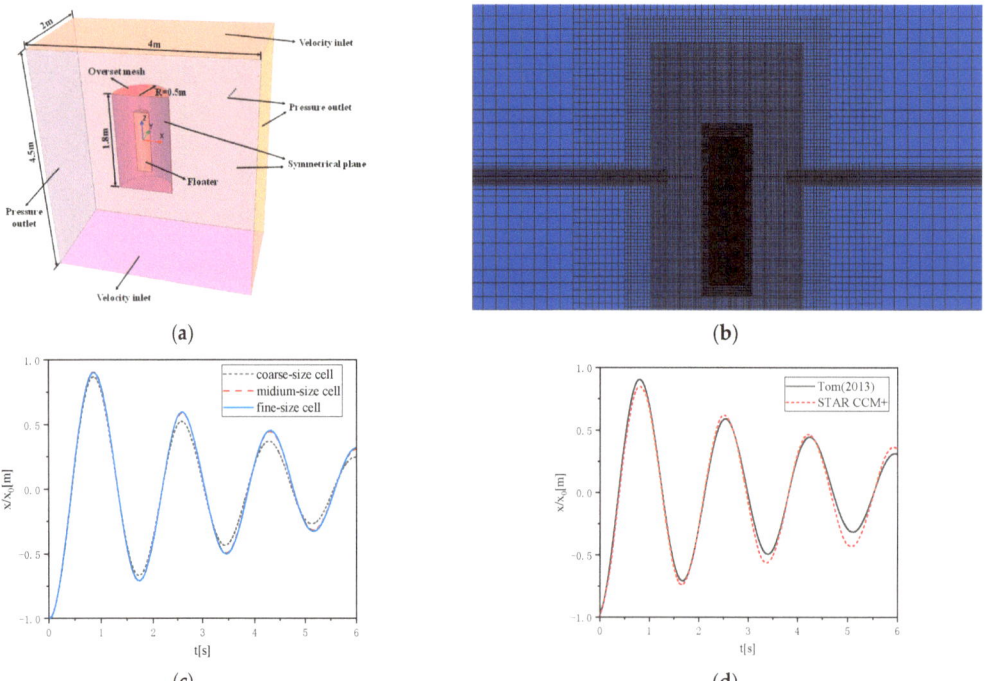

Figure 2. Validation of the numerical calculation results of STAR CCM+: (**a**) calculation domain and boundary conditions in STAR CCM+; (**b**) mesh generation in STAR CCM+; (**c**) comparison of simulation results of cells with different sizes; (**d**) comparison between numerical calculation and experimental data [21].

To calculate the response of more accurately, it is necessary to consider the viscous effect in ANSYS-AQWA. Simplify the floater into a cylinder with a draft of 3 m, and then CFD-free decay tests are carried out for four groups of floaters with different radii, and the viscous damping of the floaters is calculated according to Equations (2)–(4). Relevant parameters are shown in Table 2.

Table 2. Basic parameters for viscosity correction of floaters with different radii.

Radius (m)	Quality (Kg)	Hydrostatic Stiffness (N/m)	Viscous Damping (Ns/m)	Damping Correction Coefficient
2	3.76×10^4	1.26×10^5	0.38×10^4	0.0229
3	8.46×10^4	2.83×10^5	1.15×10^4	0.0291
4	1.50×10^5	5.05×10^5	2.02×10^4	0.0272
5	2.35×10^5	7.89×10^5	2.97×10^4	0.0244

The calculated viscous damping of floaters with different radii is introduced into ANSYS-AQWA as a linear floater damping term, and then the free attenuation motion of floaters with different radii is calculated, and the motion results under the inviscid correction are compared with the STAR CCM+ calculation results, such as Figure 3. The free attenuation motion characteristics of the revised floaters with different radii are basically consistent with the CFD method, which verifies the accuracy of the viscosity correction results.

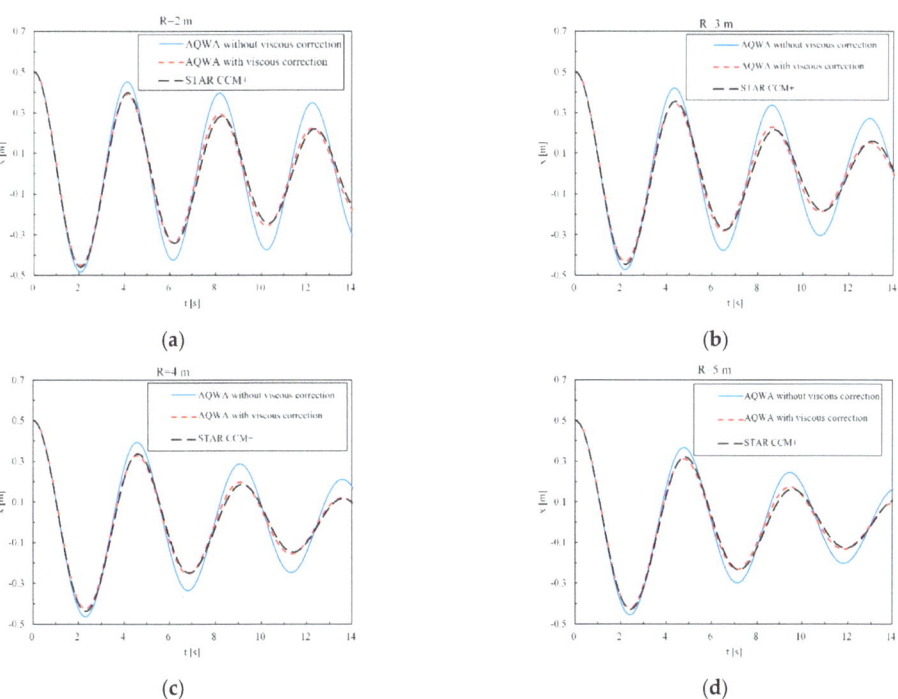

Figure 3. Attenuation motion of floaters with different radii before and after viscosity correction compared with STAR CCM+: (**a**) R = 2 m; (**b**) R = 3 m; (**c**) R = 4 m; (**d**) R = 5 m.

3.2. Floater RAO in Free-Floating State

For the point absorber WEC device, the heave RAO of the floater is a key index to evaluate the power generation efficiency of the floater. In this study, four groups of floaters with different radii are respectively combined with the semi-submersible platform foundation, and five wave directions of 0, 45, 90, 135 and 180 are selected, such as shown in Figure 4a, taking 0 wave direction as an example, the heave RAO values of the floater and the platform under different radii are compared, as shown in Figure 4b, the semi-submersible platform foundation RAO is almost the same under different floater radii, so only the platform RAO curve with a radius of 2 m is displayed. The platform motion amplitude and resonance frequency are smaller than those of the floater, and the peak period of RAO is around 20 s, exceeding the normal wave period. When the radius of the floater is 2 m in the platform, the amplitude of heave RAO is the largest and decreases with the increase of floater radius.

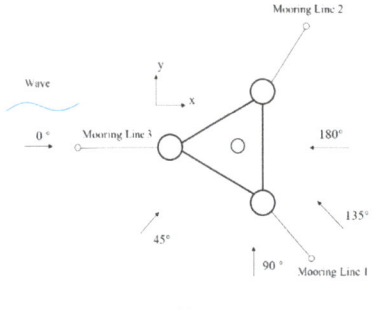
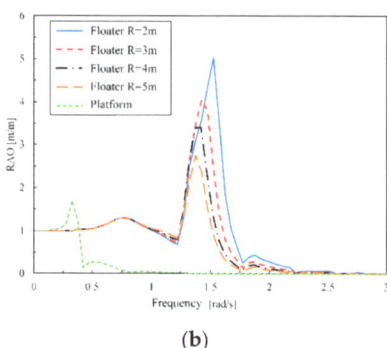

(a) (b)

Figure 4. Comparison of device layout with floater RAO with different radii: (a) overall mooring layout; (b) RAO of floater with different radii floating freely in 0 wave direction.

To explore the performance improvement of different-size floaters in the combined platform under different wave directions, the vertical heave RAO of four groups of size floaters in different wave directions was compared with that of a single floater, such as Figure 5 shown. When the radius of the floater is 2 m and 3 m, the floater RAO in the combined power generation platform is higher than that of a single floater at 0 and 180 degrees, and it is lower at 45 and 90 degrees, but it is almost the same as that of a single floater at 135 degrees. When the radius of the floater is 4 m, the floaters of the combined platform with five different waves are all higher than that of a single floater, while when the radius is 5 m, the RAO value of the heave of only 135 wave direction is slightly lower than that of a single floater. Among them, the four kinds of radius floaters under the combined platform have the best performance under 0 wave direction, which is consistent with the results in the literature [17], so 0 wave direction is taken as the subsequent calculation condition.

3.3. RAO and Average Power of the Floater in Free-Floating State

ANSYS-AQWA is used to calculate the hydrodynamic coefficient, considering the coupling effect of the PTO system between the floater and the semi-submersible platform foundation, choosing K_{PTO} = 10 kN and B_{PTO} = 20 kN·s/m, the frequency-domain motion amplitude of the floater's actual power generation can be calculated according to Equation (5), and compared with the motion amplitude and power generation of a single point absorber under the same PTO system, the RAO values under the radius of four groups of floaters and the average power generation in frequency-domain can be obtained, such as Figures 6 and 7 shown. The result shows that under the four groups of floater radii when the RAO value of the floater in the united platform is at a lower frequency, a higher peak value appears because it is close to the resonance frequency of the platform foundation, but

the corresponding power generation at this frequency is still low, and the maximum power generation is obtained when the frequency is 1.37 rad/s. At the same time, compared with a single floater, the floater RAO and power generation of the combined power generation platform with four groups of floater radii are significantly improved compared with a single floater.

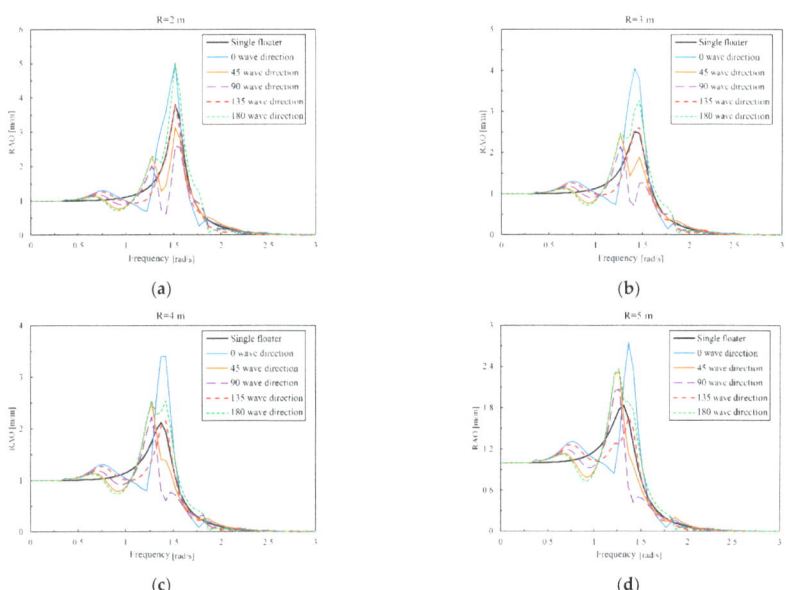

Figure 5. Comparison of RAO with different waves floater: (**a**) R = 2 m; (**b**) R = 3 m; (**c**) R = 4 m; (**d**) R = 5 m.

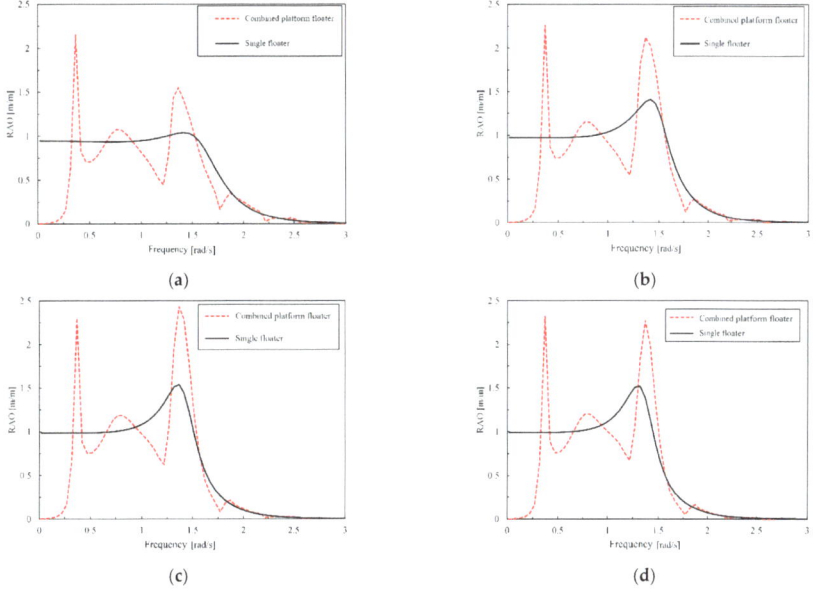

Figure 6. Comparison of RAO under different floater radii: (**a**) R = 2 m; (**b**) R = 3 m; (**c**) R = 4 m; (**d**) R = 5 m.

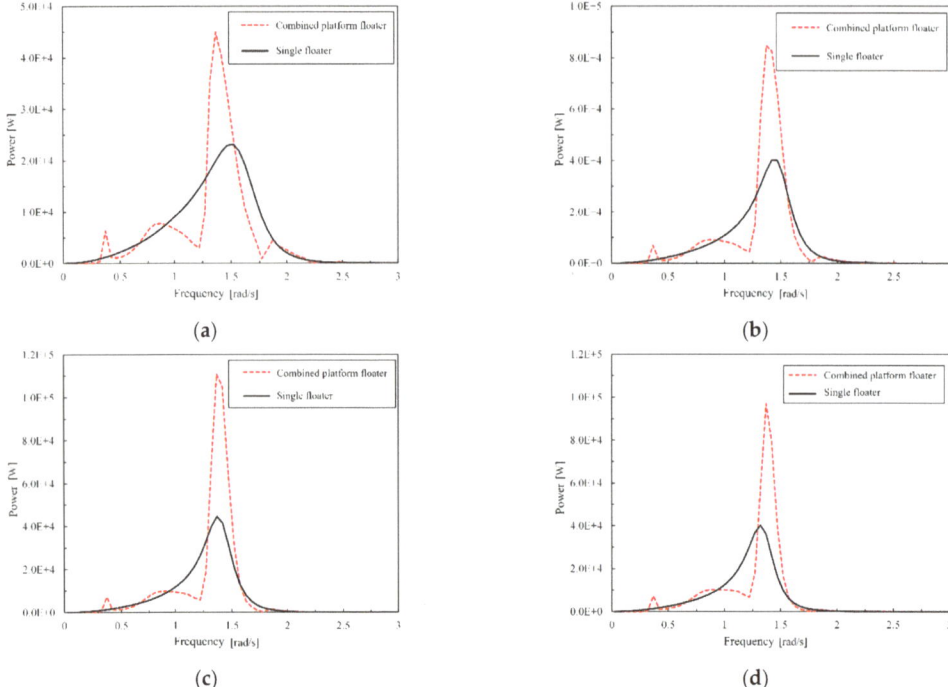

Figure 7. Comparison of power generation under different floater radii: (**a**) $R = 2$ m; (**b**) $R = 3$ m; (**c**) $R = 4$ m; (**d**) $R = 5$ m.

4. Response Analysis in Time-Domain

To further explore the influence of different sizes of floaters on the overall performance of the combined power generation platform, ANSYS-AQWA is used to further carry out time-domain analysis and analyzes the response and power generation efficiency of point absorbers in the combined power generation platform from two aspects of regular waves and irregular waves, respectively. In ANSYS-AQWA, the Fender module is used to establish a connection between the joint platform and the point absorber to simulate the PTO system, and external forces are applied to the floater and the platform to keep the initial balance, such as shown in Figure 8a, PTO parameters are also selected as $K_{PTO}= 10$ kN and $B_{PTO}= 20$ kN·s/m, To verify the accuracy of time-domain simulation, taking a floater with a radius of 4 m as an example, the average power at different calculation frequencies in time-domain and the generated power in frequency-domain are compared and analyzed under the condition of only considering heave motion, such as shown in Figure 8b, the comparison results are basically consistent, which verifies the accuracy of the analysis method. On this basis, further consider the mooring system of the semi-submersible platform foundation and the mooring parameters reference [19,20], as shown in Table 3. Therefore, the response and power generation efficiency under the condition of regular wave and irregular wave are analyzed, respectively.

4.1. Analysis of Dynamic Response and Wave Energy Conversion Efficiency of Combined Platform under Regular Waves

According to the four groups of floater radii, to verify whether the generated power of the United platform reaches the maximum value at the resonance frequency of 1.37 rad/s, firstly, under the condition of 0 wave direction regular wave with the amplitude of 1 m, the floaters of the combined platform are analyzed at 1.22 rad/s, 1.27 rad/s, 1.32 rad/s,

1.37 rad/s, 1.42 rad/s and 1.47 rad. The floaters in the combined platform all reach the maximum motion amplitude when the frequency is 1.37 rad/s, which proves the accuracy of calculating the resonance frequency in the frequency-domain.

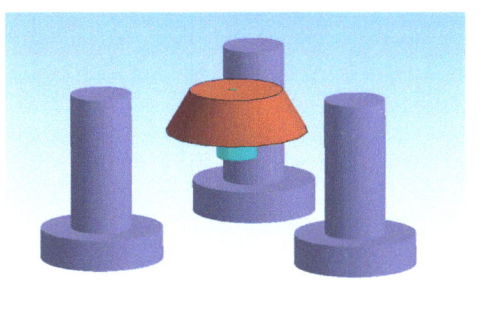
(a)

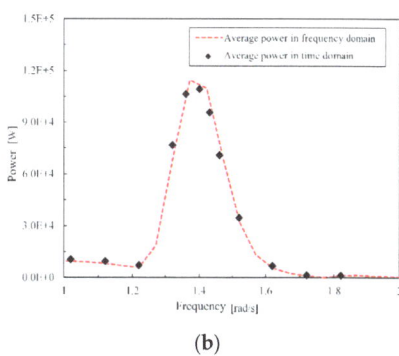
(b)

Figure 8. Frequency-domain vs. time-domain verification: (**a**) ANSYS-AQWA time-domain simulation; (**b**) Frequency-domain and time-domain power comparison.

Table 3. Mooring System Properties [19,20].

Geometric Parameter	Value
Number of Mooring Lines	3
Angle Between Adjacent Lines	120°
Depth to Anchors Below SWL	200 m
Depth to Fairleads Below SWL	14 m
Radius to Anchors from Platform Centerline	837.6 m
Radius to Fairleads from Platform Centerline	40.868 m
Unstretched Mooring Line Length	835.5
Mooring Line Diameter	0.0766 m
Equivalent Mooring Line Mass Density	113.35 kg/m
Equivalent Mooring Line Mass in Water	108.63 kg/m
Equivalent Mooring Line Extensional Stiffness	753.6 MN
Hydrodynamic Drag Coefficient for Mooring Lines	1.1
Hydrodynamic Added-Mass Coefficient for Mooring Lines	1.0
Seabed Drag Coefficient for Mooring Lines	1.0
Structural Damping of Mooring Lines	2.0%

Based on the above conclusions, the responses of the floater and the single floater in the combined power generation platform at the resonance frequency of 1.37 rad/s are further compared and analyzed. As shown in Figure 9, the displacement of the floater in the combined power generation platform is larger than that of the single floater under four groups of floater radii, which verifies the results of the frequency-domain analysis. At the same time, when the radius is 4 m, the response of a single floater and combined power generation floater is greater than that of other floaters. Then, the velocity and power generation in the combined platform with different floater radii are further compared. As shown in Figures 10 and 11, in this PTO system, the response and power generation first increase and then decrease with the floater radius. The maximum value is 1.25×10^5 W when the radius is 4 m. Meanwhile, the power generation of a single floater also increases

first and then decreases with the radius of the floater, and the maximum is 4.57×10^4 W when the radius is 4 m. Meanwhile, the power generation of the floater in the combined power generation platform is significantly higher than that of a single floater with the same radius.

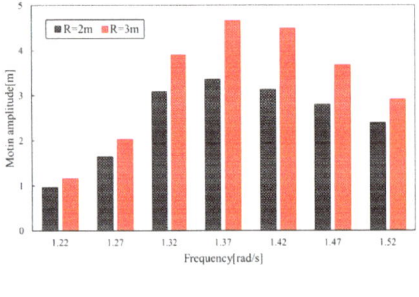

(a)

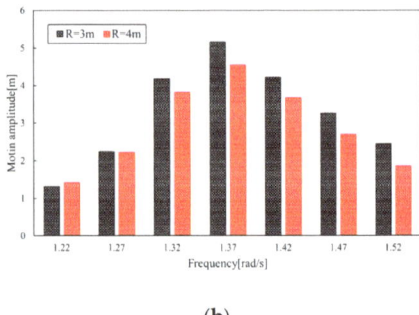

(b)

Figure 9. Motion amplitude of four groups of floater radii under regular waves with different frequencies: (**a**) R = 2 m & R = 3 m; (**b**) R = 4 m & R = 5 m.

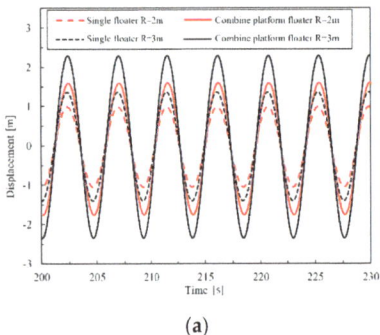

(a)

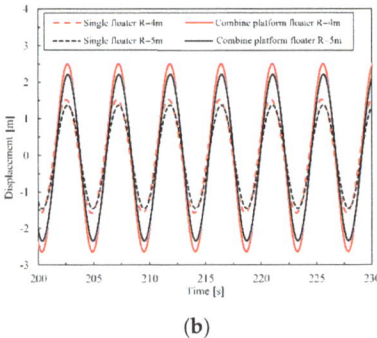

(b)

Figure 10. Displacement comparison of floaters with different sizes under regular waves: (**a**) $R = 2$ m & $R = 3$ m; (**b**) $R = 4$ m & $R = 5$ m.

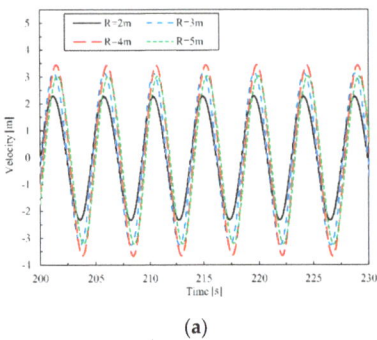

(a)

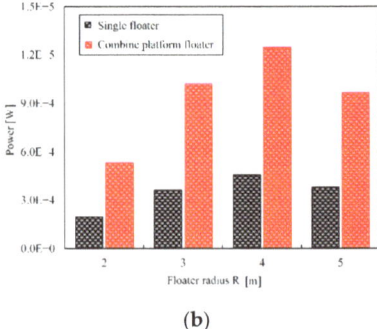

(b)

Figure 11. Comparison of velocity and power generation of floaters with different sizes under regular waves: (**a**) floater velocity comparison; (**b**) power comparison.

4.2. Analysis of Dynamic Response and Wave Energy Conversion Efficiency of Combined Platform under Irregular Waves

To more truly simulate the power generation performance of the combined power generation platform under the actual sea-states and further consider the irregular wave conditions, the JONSWAP wave spectrum is selected by its universality, with the meaningful wave height of 1 m, spectral peak factor of 3.3 and the wave direction of 0, and the corresponding frequencies of the peak periods are calculated under five working conditions of 1.27 rad/s, 1.32 rad/s, 1.37 rad/s, 1.42 rad/s and 1.47 rad/s, respectively. Three-hour Z-direction wave surface elevation velocity chronogram based on the JONSWAP spectrum is shown in Figure 12. Among them, when the radius is 4 m, the comparison of floater speed and power generation in the combined power generation platform is shown in Figure 13. When the frequency corresponding to the peak period is 1.37 rad/s, the movement speed and power generation of the floating platform are significantly improved compared with other frequencies. Therefore, it can be considered that the wind-wave combined platform proposed in this study can reach the maximum power generation when the wave frequency is 1.37 rad/s.

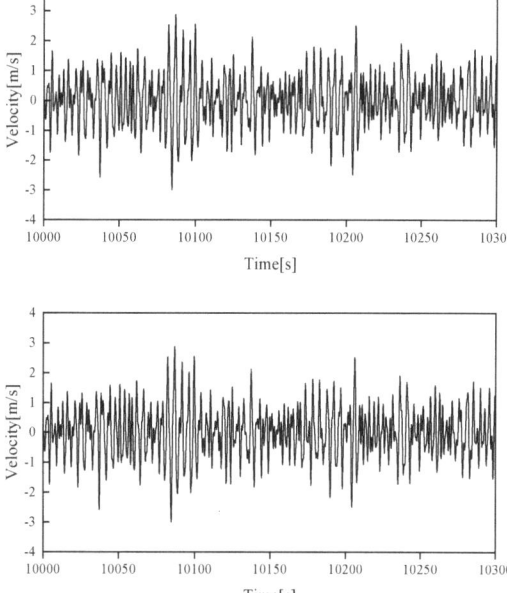

Figure 12. Three-hour Z-direction wave surface elevation velocity chronogram based on the JONSWAP spectrum.

Further, the frequency corresponding to the best peak period of 1.37 rad/s is selected to compare the power generation of the floater with that of a single floater in the combined power generation platform, as shown in Figure 14. The movement speed of the floater of the combined platform is higher than that of the single floater, and the average power generation of four groups of floaters with a radius of 4 m can also reach the maximum power generation, which is also significantly higher than that of the single floater. This further verifies the results of regular wave time-domain.

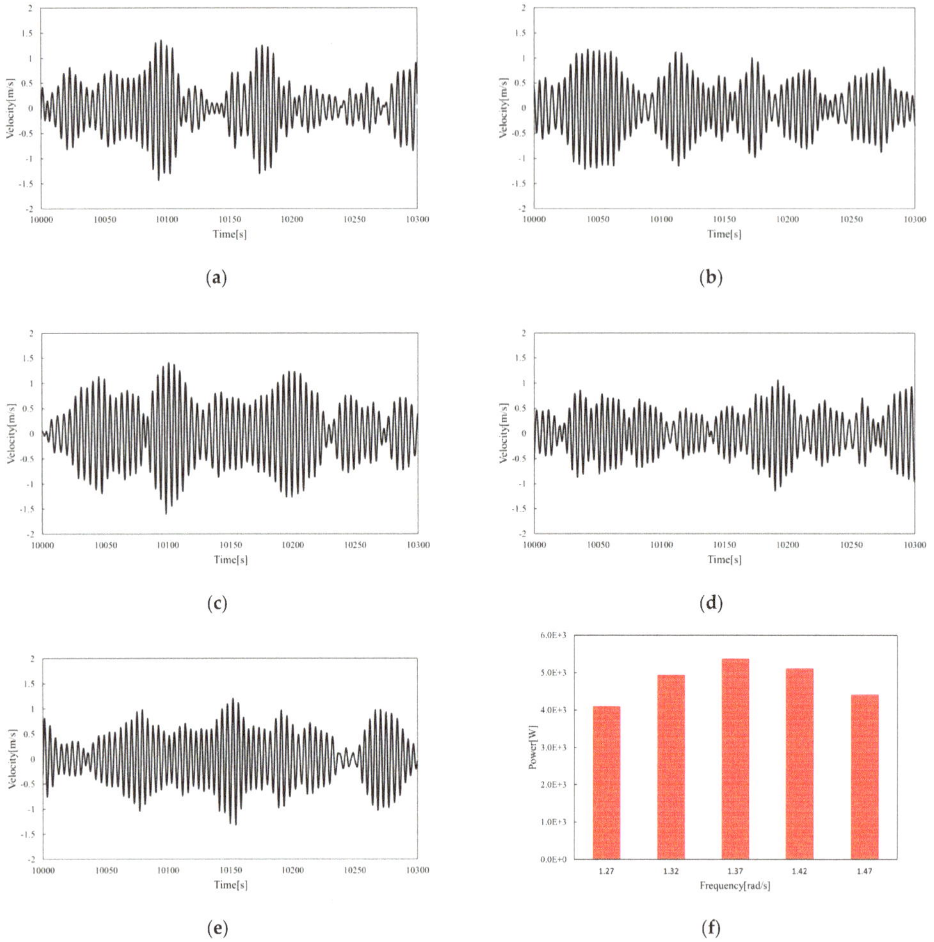

Figure 13. Comparison of velocity and power generation of floaters with different sizes under irregular waves: (**a**) ω = 1.27 rad/s; (**b**) ω = 1.32 rad/s; (**c**) ω = 1.37 rad/s; (**d**) ω = 1.42 rad/s; (**e**) ω = 1.47 rad/s; (**f**) power comparison.

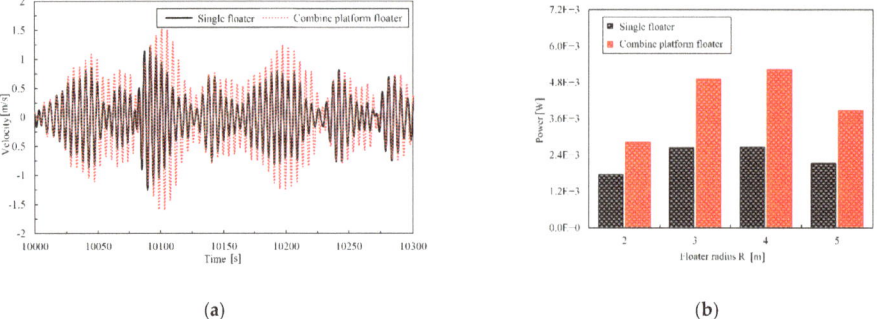

Figure 14. Comparison of velocity and power generation of floaters with different sizes under irregular waves: (**a**) floater velocity comparison; (**b**) power comparison.

5. Conclusions

In this study, the semi-submersible foundation is combined with the floater WEC to realize the combined utilization of wind energy and wave energy. The free decay test is carried out by STAR CCM+, and the frequency-domain calculation results of ANSYS-AQWA are effectively corrected. By comparing five typical wave directions and different wave frequencies, the optimal sea states of the combined device for WEC are determined. By analyzing the RAO of floaters with different radii and calculating the power generation, the optimal structure size of floaters is determined. According to the data analysis, the following conclusions are drawn as follows:

1. The viscous damping of the point absorber has a great influence on the response of the floater itself. The viscous correction method adopted in this study provides linear damping, which makes up for the ignorance of the viscous effect of fluid in ANSYS-AQWA.
2. When the wave direction is $0°$, and the wave frequency is 1.37 rad/s, the floating wind-wave combined platform achieves the highest power generation efficiency among the sea-stated considered in this study.
3. Under the configuration of PTO stiffness and damping in this research, the point absorber with a radius of 4 m has the most obvious effect on improving the WEC efficiency of the combined plant.
4. Compared with the single point absorber, the floating wind-wave combined platform proposed in this study has a significant increase in power generation, and the foundation response of the semi-submersible platform is not affected by the point absorber, which is of great significance for improving the overall power generation efficiency and reducing the combined power generation cost.

However, there are still many shortcomings in this study. This study only aims at the size design and optimization of the oscillating floating-point absorber. In a further study, WECs such as oscillating water column type or overtopping type will be considered, and the difference in power generation of different WECs will be compared. At the same time, to improve the overall power generation of the combined platform, the array combination of several floaters to realize high-power output will also be studied in the follow-up work. Since this study does not cut down the unreal wave elevation due to the suspending standing waves' effect, the following work will consider including viscosity correction to the semi-submersible foundation based on the method provided by Chen et al. [27] and Liu et al. [28], which will make the coupling analysis result of the combined power generation platform more reasonable. In addition, the survivability of the combined power generation platform under extreme sea states will also be studied.

Author Contributions: X.Z.: methodology, software, investigation, data curation, writing—original draft. B.L.: methodology, software, investigation, data curation. Z.H.: data curation, investigation, and software. J.D.: data curation, investigation, and software. P.X.: writing—review and editing. M.C.: supervision, writing—review and editing, and funding acquisition. All authors have read and agreed to the published version of the manuscript.

Funding: This research was funded by the National Natural Science Foundation of China, grant number 52171275. and the Natural Science Foundation of Hainan Province, China, grant number 520MS072.

Conflicts of Interest: The authors declare no conflict of interest.

References

1. Peiffer, A.; Roddier, D.; Aubault, A. Design of a point absorber inside the WindFloat structure. In Proceedings of the International Conference on Offshore Mechanics and Arctic Engineering, Rotterdam, The Netherlands, 19–24 June 2011; Volume 44373, pp. 247–255.
2. Aubault, A.; Alves, M.; Sarmento, A.; Roddier, D.; Peiffer, A. Modeling of an oscillating water column on the floating foundation WindFloat. In Proceedings of the International Conference on Offshore Mechanics and Arctic Engineering, Rotterdam, The Netherlands, 19–24 June 2011; Volume 44373, pp. 235–246.

3. Soulard, T.; Babarit, A. Numerical assessment of the mean power production of a combined wind and wave energy platform. In Proceedings of the International Conference on Ocean, Offshore and Arctic Engineering (OMAE), Rio de Janeiro, Brazil, 1–6 July 2012; Volume 44946, pp. 413–423.
4. Muliawan, M.J.; Karimirad, M.; Moan, T.; Gao, G. STC (Spar-Torus Combination): A combined spar-type floating wind turbine and large point absorber floating wave energy convertor—Promising and challenging. In Proceedings of the International Conference on Offshore Mechanics and Arctic Engineering, Rotterdam, The Netherlands, 19–24 June 2011; American Society of Mechanical Engineers: New York, NY, USA, 2012; Volume 44946, pp. 667–676.
5. Luan, C.; Michailides, C.; Gao, Z.; Moan, T. Modeling and analysis of a 5 MW semi-submersible wind turbine combined with three flap-type wave energy converters. In Proceedings of the International Conference on Offshore Mechanics and Arctic Engineering, San Francisco, CA, USA, 8–13 June 2014; American Society of Mechanical Engineers: New York, NY, USA; Volume 45547, p. V09BT09A028.
6. Gaspar, J.F.; Kamarlouei, M.; Thiebaut, F.; Guedes Soares, C. Compensation of a hybrid platform dynamics using wave energy converters in different sea state conditions. *Renew. Energy* **2021**, *177*, 871–883. [CrossRef]
7. Michailides, C.; Luan, C.; Gao, Z.; Moan, T. Effect of flap type wave energy converters on the response of a semi-submersible wind turbine in operational conditions. In Proceedings of the International Conference on Offshore Mechanics and Arctic Engineering, San Francisco, CA, USA, 8–13 June 2014; Volume 45547, p. V09BT09A014.
8. Wan, L.; Gao, Z.; Moan, T.; Lugni, C. Comparative experimental study of the survivability of a combined wind and wave energy converter in two testing facilities. *Ocean. Eng.* **2016**, *111*, 82–94. [CrossRef]
9. Muliawan, M.J.; Karimirad, M.; Gao, Z.; Moan, T. Extreme responses of a combined spar-type floating wind turbine and floating wave energy converter (STC) system with survival modes. *Ocean. Eng.* **2013**, *65*, 71–82. [CrossRef]
10. Hallak, T.S.; Karmakar, D.; Guedes Soares, C. Hydrodynamic performance of semi-submersible FOWT combined with point-absorber WECs. In *Developments in Maritime Technology and Engineering*; CRC Press: London, UK, 2021; pp. 577–585.
11. Wang, Y.; Zhang, L.; Michailides, C.; Wan, L.; Shi, W. Hydrodynamic Response of a Combined Wind–Wave Marine Energy Structure. *J. Mar. Sci. Eng.* **2020**, *8*, 253. [CrossRef]
12. González, I.T.; Ricci, P.; Lara, M.J.S.; Morán, G.P.; Papo, F.B. Design, modelling and analysis of a combined semi-submersible floating wind turbine and wave energy point-absorber. In Proceedings of the ASME 2013 32st International Conference on Ocean, Offshore and Arctic Engineering, Nantes, France, 9–15 June 2013; pp. 1–11.
13. Sun, K.; Yi, Y.; Zheng, X.; Cui, L.; Zhao, C.; Liu, M.; Rao, X. Experimental investigation of semi-submersible platform combined with point-absorber array. *Energy Convers. Manag.* **2021**, *245*, 114623. [CrossRef]
14. Hu, J.; Zhou, B.; Vogel, C.; Liu, P.; Willden, R.; Sun, K.; Zang, J.; Geng, J.; Jin, P.; Cui, L.; et al. Optimal design and performance analysis of a hybrid system combing a floating wind platform and wave energy convertors. *Appl. Energy* **2020**, *269*, 114998. [CrossRef]
15. Lee, H.; Poguluri, S.K.; Bae, Y.H. Performance analysis of multiple wave energy convertors placed on a floating platform in the frequency-domain. *Energies* **2018**, *11*, 406. [CrossRef]
16. Hantoro, R.; Septyaningrum, E.; Hudaya, Y.R.; Utama, I.K.A.P. Stability analysis for trimaran pontoon array in wave energy converter–pendulum system (WEC-PS). *Brodogr. Teor. I Praksa Brodogr. I Pomor. Teh.* **2022**, *73*, 59–68. [CrossRef]
17. Chen, M.; Wang, R.; Xiao, P.; Zhu, L.; Li, F.; Sun, L. Numerical analysis of a floating semi-submersible wind turbine integrated with a point absorber wave energy convertor. In Proceedings of the The 30th International Ocean and Polar Engineering Conference, Virtual, 12 October 2020.
18. Chen, M.; Xiao, P.; Zhou, H.; Li, C.B.; Zhang, X. Fully Coupled Analysis of an Integrated Floating Wind-Wave Power Generation Platform in Operational Sea-states. *Front. Energy Res.* **2022**, *10*, 931057. [CrossRef]
19. Robertson, A.; Jonkman, J.; Masciola, M.; Song, H.; Goupee, A.; Coulling, A.; Luan, C. *Definition of the Semisubmersible Floating System for Phase II of OC4*; NREL/TP-5000-60601; National Renewable Energy Lab. (NREL): Golden, CO, USA, 2014.
20. Cheng, P.; Huang, Y.; Wan, D. A numerical model for fully coupled aero-hydrodynamic analysis of floating offshore wind turbine. *Ocean. Eng.* **2019**, *173*, 183–196. [CrossRef]
21. Tom, N.M. *Design and Control of a Floating Wave-Energy Convertor Utilizing a Permanent Magnet Linear Generator*; UC Berkeley: Berkeley, CA, USA, 2013.
22. Sun, K.; Xie, G.; Zhou, B. Type selection and hydrodynamic performance analysis of wave energy convertors. *J. Harbin Eng. Univ.* **2021**, *42*, 8–14.
23. Cummins, W.E. *The Impulse Response Function and Ship Motions*; David Taylor Model Basin Washington DC: Washington, DC, USA, 1962.
24. Ogilvie, T.F. Recent progress toward the understanding and prediction of ship motions. In Proceedings of the 5th ONR Symp. on Naval Hydrodynamics, Bergen, Norway, 10–12 September 1964.
25. Folley, M. *Numerical Modelling of Wave Energy Converters: State-of-the-Art Techniques for Single Devices and Arrays*; Academic Press: London, UK, 2016; Volume 23.
26. Chen, M.; Xiao, P.; Zhang, Z.; Sun, L.; Li, F. Effects of the end-stop mechanism on the nonlinear dynamics and power generation of a point absorber in regular waves. *Ocean. Eng.* **2021**, *242*, 110123. [CrossRef]
27. Chen, M.; Guo, H.; Wang, R.; Tao, R.; Cheng, N. Effects of gap resonance on the hydrodynamics and dynamics of a multi-module floating system with narrow gaps. *J. Mar. Sci. Eng.* **2021**, *9*, 1256. [CrossRef]
28. Liu, H.; Chen, M.; Han, Z.; Zhou, H.; Li, L. Feasibility Study of a Novel Open Ocean Aquaculture Ship Integrating with a Wind Turbine and an Internal Turret Mooring System. *J. Mar. Sci. Eng.* **2022**, *10*, 1729. [CrossRef]

Article

Modeling, Experimental Analysis, and Optimized Control of an Ocean Wave Energy Conversion System in the Yellow Sea near Lianyungang Port

Zhongxian Chen [1,2,*], Xu Li [1], Yingjie Cui [1] and Liwei Hong [3]

1. School of Intelligence Manufacturing, Huanghuai University, Zhumadian 463000, China
2. Henan Key Laboratory of Smart Lighting, Huanghuai University, Zhumadian 463000, China
3. State Grid Langfang Electric Power Supply Company, Langfang 065000, China
* Correspondence: chenzhongxian@huanghuai.edu.cn

Abstract: In this paper, an ocean wave energy conversion system (OWECS) is modeled and experimented in the Yellow Sea near Lianyungang port, and an optimized control method based on the sliding mode control is proposed to improve the efficiency of OWECS. Firstly, a motion model of a double-buoy OWECS is presented using a complex representation method, and the analysis results indicate that the efficiency of converting ocean wave energy into the outer buoy's mechanical power is highest in a suitable ocean wave period. Secondly, a double-buoy OWECS is constructed and experimented in the Yellow Sea near Lianyungang port, which verified the correctness of the above analysis results. Lastly, in order to further improve the efficiency of the double-buoy OWECS, a sliding mode control method based on a linear generator is proposed to realize the phase synchronization between the outer buoy and ocean waves, and the simulation results may be beneficial for the next ocean test of the double-buoy OWECS.

Keywords: motion model; ocean wave energy; buoy; efficiency; optimize control

Citation: Chen, Z.; Li, X.; Cui, Y.; Hong, L. Modeling, Experimental Analysis, and Optimized Control of an Ocean Wave Energy Conversion System in the Yellow Sea near Lianyungang Port. *Energies* **2022**, *15*, 8788. https://doi.org/10.3390/en15238788

Academic Editors: Yan Bao, Guanghua He and Liang Sun

Received: 2 November 2022
Accepted: 20 November 2022
Published: 22 November 2022

Publisher's Note: MDPI stays neutral with regard to jurisdictional claims in published maps and institutional affiliations.

Copyright: © 2022 by the authors. Licensee MDPI, Basel, Switzerland. This article is an open access article distributed under the terms and conditions of the Creative Commons Attribution (CC BY) license (https://creativecommons.org/licenses/by/4.0/).

1. Introduction

As a kind of renewable clean energy, ocean wave energy has attracted more and more attention in the world [1,2]. During the past 50 years, ocean wave energy conversion has been investigated and developed on a large scale, and various prototypes have been tested in the laboratory or real ocean waves [3–5]. According to the system structure, an OWECS can be classified as near-shore and offshore [6]. Regardless of the type of OWECS, buoys (or float balls) and generators are necessary.

A buoy (or float ball) is an energy transfer device that converts ocean wave energy into mechanical energy, and then drives a generator to output electrical energy. As shown in Figure 1, under the action of ocean waves, the floating ball can move in the vertical direction, which drives the linear generation to convert ocean wave energy into electrical energy. In the past 5 years, novel kinds of OWECS have been proposed and researched. For example, a hybrid OWECS which contains an oscillating water column (OWC) and oscillating buoy (OB) was proposed by Saishuai et al., and the research results indicated that the hybrid OWECS provided higher energy conversion, as well as better wave attenuation performance, compared with the isolated OWC and OB devices [7]. In addition, an oscillating buoy (OB) single-pontoon floating breakwater (SPFB) and an oscillating water column (OWC) dual-pontoon floating breakwater (DPFB) were evaluated and compared. The comparison results showed that the maximum conversion efficiency of the OWC for the optimal opening ratio was higher than that of the OB [8]. Furthermore, some other kinds of OWECS were researched in the last 5 years, such as the point-absorbing wave energy converter (WEC) with the new mechanism structure of zero pressure angle, the

smart floatable open-water wave energy converter, and the variable-geometry wave energy converter [9–11].

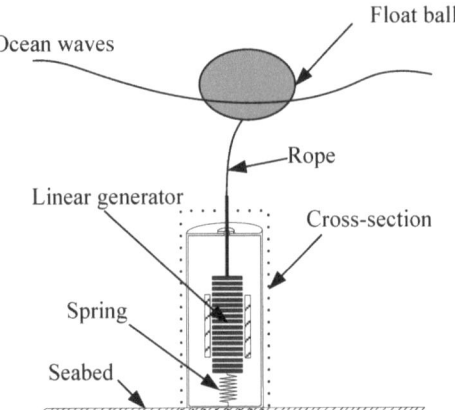

Figure 1. Structure of an ocean wave energy conversion device.

Actually, the motion of buoys (or float balls) plays an important role in the operation efficiency of an OWECS. Only when the natural frequency of buoys (or float balls) is identical to the ocean wave's frequency can the efficiency of ocean wave energy converting into electrical energy be maximized [12,13]. Therefore, the design and modeling of buoys (or float balls) is one of the important topics in the research of OWECS.

Furthermore, a generator is an energy conversion device that converts mechanical energy into electrical energy. In order to improve the operational efficiency of a generator, some novel generators have been proposed, such as the superconducting magnet-excited linear generator, linear switched reluctance generator, and Halbach magnetized linear permanent magnet generator [14–16]. Compared with conventional rotary generators, the high efficiency and simple structure make the linear generator an attractive candidate for OWECS. However, due to the variable nature of wave periods and wave heights of ocean waves, the non-controlled generator does not play a key role in the operational efficiency of OWECS [17].

In this paper, according to the wave periods of the Yellow Sea near Lianyungang port, the model test of a double-buoy OWECS was carried out in the Yellow Sea near Lianyungang port. Firstly, the motion model of a double-buoy OWECS is presented and analyzed. Secondly, a double-buoy OWECS is experimented in the Yellow Sea near Lianyungang port, proving the correctness of the analysis results of the motion model of the double-buoy OWECS. Lastly, in order to further improve the efficiency of the double-buoy OWECS, a sliding mode control method based on the linear generator is proposed and simulated in Section 4 of this paper. At the end of this paper, the principal restrictions of study in this paper are discussed.

2. Motion Model of Double-Buoy OWECS

2.1. Vertical Direction Speed of Buoy

A buoy floating in ocean waves is shown in Figure 2a. Under the action of ocean waves, the buoy can be moved in the vertical direction z, surge direction x, and sway direction y. Figure 2b shows a double-buoy OWECS with an outer buoy and inner buoy. Due to the different draught of the outer buoy and inner buoy, a relative motion between them occurs, which drives the generator (installed in the upper end of inner buoy) to convert ocean wave energy into electrical energy.

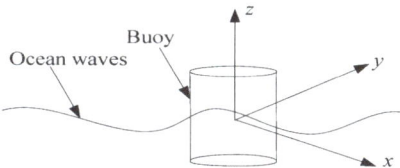

(a) The motion direction of buoy in ocean waves

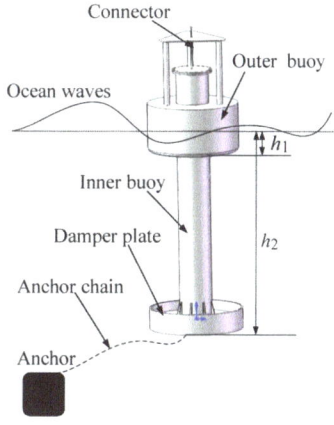

(b) Double-buoy OWECS

Figure 2. The operational principle of double-buoy OWECS.

In the vertical direction, the acceleration formula of buoy can be described as

$$m\hat{a}_z = \hat{F}_z + \hat{F}_r + \hat{F}_b + \hat{F}_f, \quad (1)$$

where m is the mass (in kg), $\hat{a}_z$ is the acceleration in the vertical direction, $\wedge$ represents a complex representation, $\hat{F}_z$ is the vertical direction ocean wave force, $\hat{F}_r$ is the radiation force from the relative motion between buoy and ocean waves, $\hat{F}_b$ is the hydrostatic buoyancy force, and $\hat{F}_f$ is the friction force.

Usually, the diameter of a buoy is smaller than the wavelength of ocean waves; thus, the method of Froude–Krylow force and small object approximation can be used [18]. The vertical direction ocean wave force $\hat{F}_z$ can be written as

$$\hat{F}_z = \left[\rho g S_w - \omega^2 \rho V (1 + \mu_z)\right] \hat{\eta}, \quad (2)$$

where ρ is the density of ocean water, g is the acceleration due to gravity, S_w is the horizontal cross-section of buoy, ω is the angular frequency of ocean waves, V is the volume of the buoy below the ocean wave surface, μ_z is the added mass coefficient of the buoy, and $\hat{\eta}$ is the wave amplitude of the ocean wave. Considering the depth z below the ocean wave surface, the wave amplitude of the ocean wave $\hat{\eta}$ can be written as

$$\hat{\eta} = Ae^{ikz}, \quad (3)$$

where A is the amplitude of the ocean wave surface, i represents the imaginary part of the complex representation, and $k = \omega^2/g$ is the angular wave number of ocean waves. From Equation (3), it can be concluded that a greater depth z below the ocean wave surface leads to a smaller amplitude of the ocean wave.

Moreover, the radiation force $\hat{F}_r$, the hydrostatic buoyancy force $\hat{F}_b$, and the friction force $\hat{F}_f$ can be written as

$$\hat{F}_r = \left(\omega^2 m_z - i\omega R_z\right)\frac{\hat{a}_z}{-\omega^2}, \tag{4}$$

$$\hat{F}_b = -\rho g S_w \frac{\hat{a}_z}{-\omega^2}, \tag{5}$$

$$\hat{F}_f = -i\omega R_f \frac{\hat{a}_z}{-\omega^2}, \tag{6}$$

where m_z and R_z are the added mass and damping coefficient of the buoy, respectively, R_f is the friction resistance coefficient of the double-buoy OWECS.

According to the relationship between speed and acceleration, the vertical direction acceleration $\hat{a}_z$ can be written as

$$\hat{a}_z = i\omega \hat{v}_z, \tag{7}$$

where $\hat{v}_z$ is the vertical direction speed. Substituting Equations (2) and (4)–(7) into (1), the vertical direction speed of buoy can be described as

$$\hat{v}_z = \frac{\left[\rho g S_w - \omega^2 \rho V(1+\mu_z)\right]\hat{\eta}}{i\omega[m + m_z] + \left[R_f + R_z\right] + \frac{\rho g S_w}{i\omega}}. \tag{8}$$

For double-buoy OWECS (see Figure 2b), the basic parameters of double buoys are shown in Table 1. Because the diameters of buoys are smaller than the wavelength of ocean waves, the added mass can be approximately expressed as $m_z = 0.17\rho D^3$, where D is the diameter of buoys, and the damping coefficient R_z can be ignored. It is assumed that the ocean waves are regular; then, according to Equations (3) and (8) and the basic parameters of double buoys, the vertical direction speeds of the outer buoy and inner buoy are illustrated in Figure 3. Figure 3 indicates that a relative motion between the outer buoy and inner buoy occurs, which drives the generator (installed in the upper end of inner buoy) to convert ocean wave energy into electrical energy.

Table 1. The basic parameters of double buoys.

	Outer Diameter	2.4 m
Outer buoy	Inner diameter	1.0 m
	Height	1.8 m
	Draft (h_1)	0.771 m
Inner buoy	Outer diameter	0.83 m
	Height	7.9 m
	Draft (h_2)	6.059 m

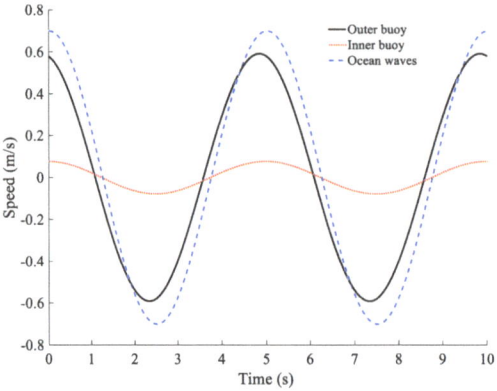

Figure 3. The vertical direction speed of the outer buoy and inner buoy.

2.2. Analysis of Energy Conversion Efficiency

According to the theory of mechanical vibration, only when the resonance condition occurs is the vertical direction displacement of the buoy maximized; thus, the energy conversion efficiency from ocean wave energy into mechanical energy is maximum [18]. For the double-buoy OWECS, the resonance condition means that the outer buoy's vertical direction speed $\hat{v}_z$ is consistent with its vertical direction ocean wave force $\hat{F}_z$ (phase difference is zero), and the inner buoy is nearly stationary. It can be concluded from Equation (8) that the angular frequency ω plays an important role in the phase difference between the vertical direction ocean wave force $\hat{F}_z$ and the vertical direction speed $\hat{v}_z$.

According to Equation (8), Table 1, and certain regular ocean waves with different period T, Figure 4 shows the outer buoy's vertical direction ocean wave force and vertical direction speed, and Table 2 lists the theoretical conversion efficiency from the ocean wave power into the outer buoy and inner buoy's mechanical power. In Table 2, the wave power is calculated as the product of wave surface power J and the buoy's horizontal cross-section S_w, and the wave surface power J can be expressed as

$$J = \frac{\rho g^2}{32\pi} T H^2, \tag{9}$$

where T is the wave period, and H is the wave height. In addition, it should be noted that the ocean wave power consists of kinetic energy (50%) and potential energy (50%), and only the kinetic energy can be converted into the buoy's mechanical power.

Table 2. The theoretical conversion efficiency from ocean wave power into the buoy's power.

Wave Amplitude (m)	Wave Period (s)	Ocean Wave (Kinetic Energy)	Outer Buoy	Inner Buoy	Efficiency
		Power (kW)			
0.7	3	10.7275	3.4206	0.5225	27.02%
0.7	5	17.879	7.0680	0.5615	36.39%
0.7	7	25.0305	6.3139	0.5258	23.12%
0.7	8	28.6065	5.9704	0.5189	19.06%

Table 2 indicates that, for the geometry of the outer buoy of a double-buoy OWECS (see Table 1), there is an optimal ocean wave period which can convert ocean wave energy into the maximum mechanical power of the outer buoy (under the natural operation state), and the optimal ocean wave period T is about 5 s.

The Yellow Sea near Lianyungang port was the installation site of the double-buoy type OWECS; accordingly, the average wave period T and average angular frequency ω in this area are listed in the Table 3 [19]. The average wave period T in the four seasons is about 5 s, which is suitable for the operation of a double-buoy type OWECS. Actually, the real ocean waves are irregular, and the efficiency of the outer buoy should be lower than that of the regular waves. However, the analysis of conversion efficiency based on the regular waves would provide some reference for the design and experimental test of double-buoy ocean wave energy conversion in real ocean waves.

Table 3. The average wave periods and frequencies in Yellow Sea near Lianyungang port.

Season	Average Wave Periods T	Average Angular Frequency ω
Spring	5–5.5 s	1.1424–1.2566 rad/s
Summer	5.5–6 s	1.0472–1.1424 rad/s
Autumn	5–5.5 s	1.1424–1.2566 rad/s
Winter	5–5.5 s	1.1424–1.2566 rad/s

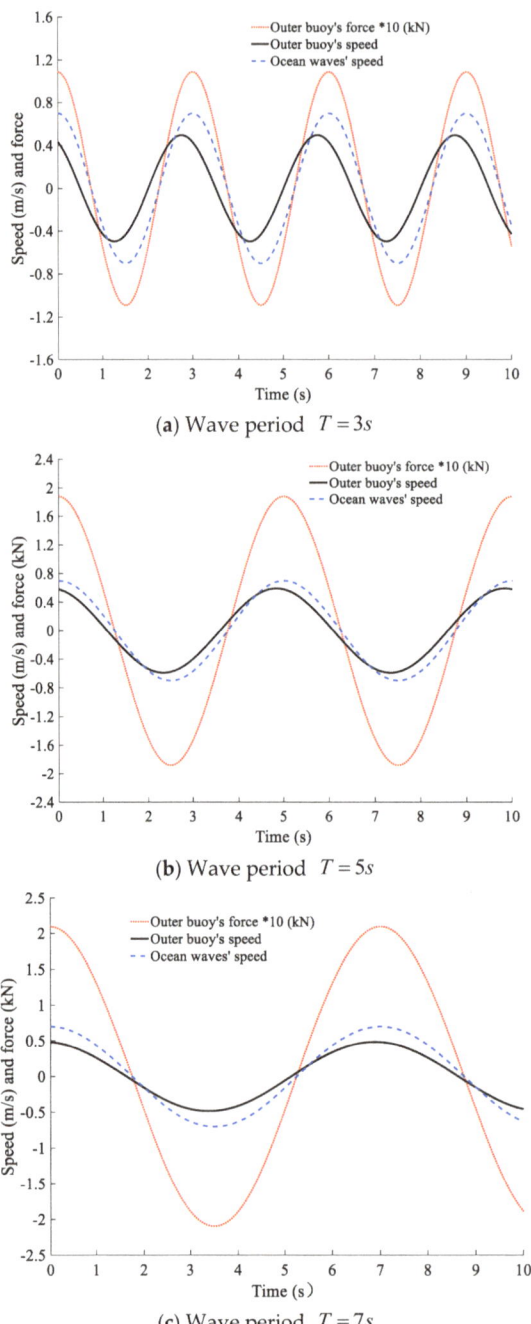

Figure 4. The vertical direction force and speed of outer buoy.

3. Experimental Results

According to the efficiency analysis of ocean wave power conversion into the outer buoy's mechanical power, a double-buoy OWECS based on the parameters of Table 1 was constructed and experimented in the Yellow Sea near Lianyungang port. The main components of the double-buoy OWECS included the outer buoy, inner buoy, linear generator, data collector, etc.

3.1. The Specific Structure of Double-Buoy OWECS

Figure 5 shows the overall structure of the linear generator, inner buoy, measuring device, outer buoy, and double-buoy OWECS. In Figure 5a, the linear generator is a permanent magnet tube, which is more suitable for use in an OWECS than a rotary generator. Figure 5b shows the internal structure of inner buoy, in which the linear generator is installed in the upper end of inner buoy. Figure 5c shows the measuring device, including the data collector, data processor, power supply, and data transmission. Figure 5d shows the overall structure of the double-buoy OWECS, with the measuring device installed in the headspace of the outer buoy.

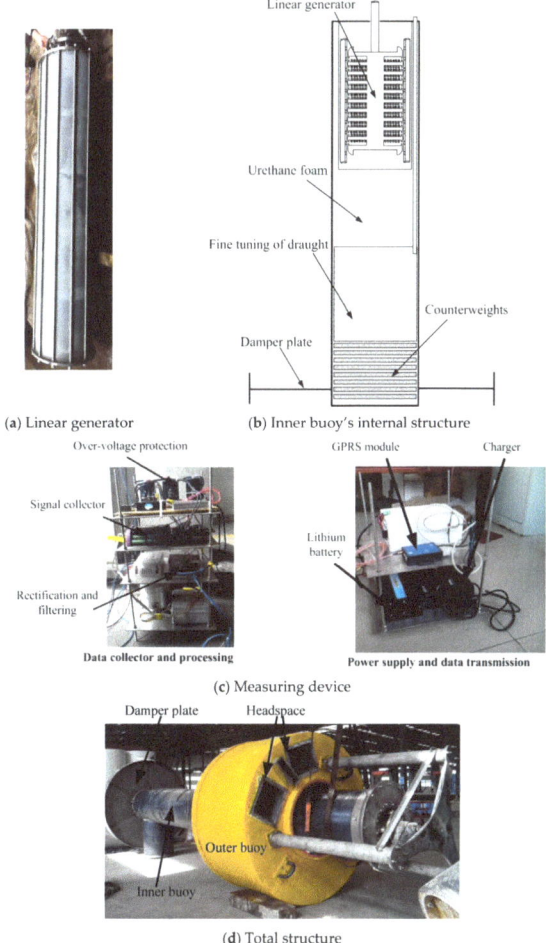

Figure 5. The specific structure of the double-buoy OWECS.

Furthermore, a diagram of the measuring device is shown in Figure 6. As shown in Figure 6, the output voltage of the linear generator (from the double-buoy OWECS) is transformed into 0–5 V by voltage regulation and sampled by the MCU (microcontroller unit). Then, the sampled data can be stored in flash memory or transmitted to the server via a GPRS module. In addition, the function of the relay is overvoltage protection.

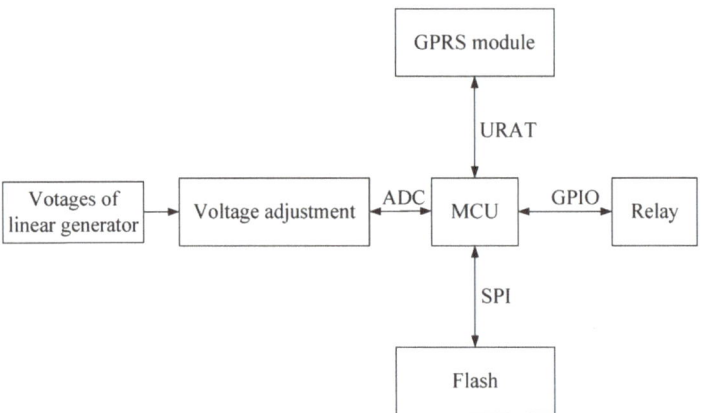

Figure 6. Diagram of measuring device.

3.2. Output Voltage and Power Analysis

Figure 7a shows the installation process of the double-buoy OWECS in the Yellow Sea near Lianyungang port, and Figure 7b shows its operation in ocean waves.

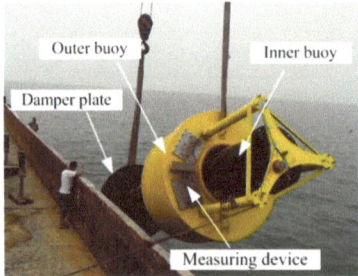

(a) Installation process of outer buoy and inner buoy

(b) Double-buoy OWECS in ocean waves

Figure 7. The installation process and operation of OWECS.

Figure 8a,b show the output load voltages of the double-buoy OWECS collected by the measuring device. Figure 8c is a power analysis of Figure 8a,b, in which the average power is 1.176 kW, and the maximum power is 4.254 kW.

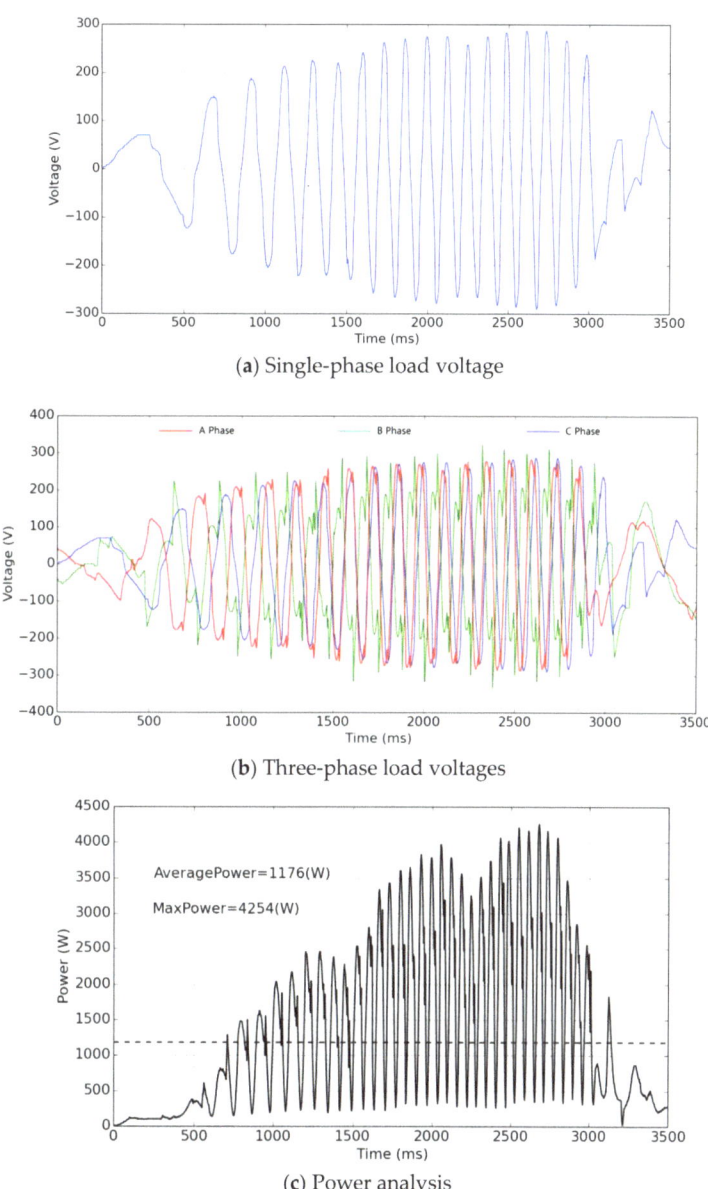

(a) Single-phase load voltage

(b) Three-phase load voltages

(c) Power analysis

Figure 8. Output load voltages and power analysis (the ocean wave period is about 5.5 s).

The output load voltages in Figure 8a,b indicate that the ocean wave period at the same time is about 5.5 s, and the ocean wave amplitude is about 0.7 m (based on the linear generator's voltage–speed characteristic). Under this assumption, according to Equation (9),

the experiment conversion efficiency from ocean wave power into electric power is 5.98% (the linear generator's output electric power divided by the ocean wave power).

In addition, if the ocean wave period is longer than 5.5 s, in the case of 8.5 s, the output load voltage of the double-buoy OWECS is as shown in Figure 9a,b, along with the average power (1.187 kW) and maximum power (2.34 kW). Furthermore, according to Equation (9), the experiment conversion efficiency from ocean wave power into electric power is 3.4%.

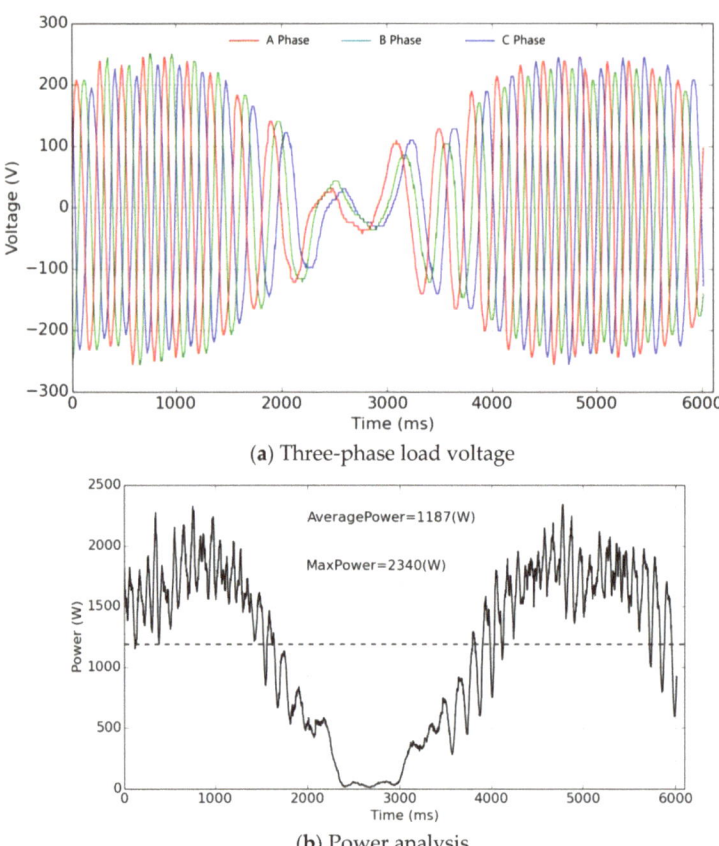

(a) Three-phase load voltage

(b) Power analysis

Figure 9. Output load voltages and power analysis (the ocean wave period is about 8.5 s).

Table 4 shows the experiment conversion efficiency from ocean wave power into electric power, when the ocean wave periods are 5.5 s and 8.5 s.

Table 4. The experiment conversion efficiency from ocean wave power into electric power.

| Wave Amplitude (m) | Wave Period (s) | Power (kW) | | Efficiency |
		Ocean Wave (Kinetic Energy)	Linear Generator	
0.7	5.5	19.667	1.176	5.98%
0.75	8.5	34.8915	1.187	3.4%

4. Efficiency Improvement of Double-Buoy OWECS

As shown in Table 4, in the natural ocean environment, the operational efficiency of the double-buoy OWECS is low. Furthermore, irregular ocean waves further reduce the operation efficiency. Therefore, considering the operation characteristics of a linear generator, an optimized control method is proposed to increase the operational efficiency of the double-buoy OWECS.

For the double-buoy OWECS, the load force of the linear generator should be considered. Therefore, the vertical direction speed of the floating buoy in the double-buoy OWECS can be written as

$$\hat{v}_z = \frac{[\rho g S_w - \omega^2 \rho V(1+\mu_z)]\hat{\eta} + \hat{F}_u}{i\omega[m+m_z] + [R_f + R_z] + \frac{\rho g S_w}{i\omega}}, \tag{10}$$

where $\hat{F}_u$ is the load force of linear generator. For the linear generator, the relationship between load force $\hat{F}_u$ and the q-axis current is described below.

4.1. The Relationship between Load Force and q-Axis Current of Linear Generator

According to the theory of electrical engineering [20], the Park transformation can be written as

$$\begin{bmatrix} s_a \\ s_b \\ s_c \end{bmatrix} = \begin{bmatrix} \cos(\theta) & -\sin(\theta) & 1 \\ \cos(\theta-120°) & -\sin(\theta-120°) & 1 \\ \cos(\theta+120°) & -\sin(\theta+120°) & 1 \end{bmatrix} \begin{bmatrix} s_d \\ s_q \\ s_0 \end{bmatrix} = Park^{-1} \begin{bmatrix} s_d \\ s_q \\ s_0 \end{bmatrix}, \tag{11}$$

where s_a is the voltage, s_b is the current, and s_c is the magnetic linkage in the form of abc coordinates. Moreover, s_d is the voltage, s_q is the current, and s_0 is the magnetic linkage in the form of $dq0$ coordinates.

In the form of abc coordinates, the active power output of the linear generator can be written as

$$\hat{P} = \hat{V}_a \hat{i}_a + \hat{V}_b \hat{i}_b + \hat{V}_c \hat{i}_c, \tag{12}$$

where $\hat{V}_a$, $\hat{V}_b$, and $\hat{V}_c$ are the voltages, and $\hat{i}_a$, $\hat{i}_b$, and $\hat{i}_c$ are the currents. Substituting Equation (11) into (12), the active power output of linear generator can be rewritten as

$$\hat{P} = \begin{bmatrix} \hat{V}_a & \hat{V}_b & \hat{V}_c \end{bmatrix} \begin{bmatrix} \hat{i}_a \\ \hat{i}_b \\ \hat{i}_c \end{bmatrix} = Park \begin{bmatrix} \hat{V}_d \\ \hat{V}_q \\ \hat{V}_0 \end{bmatrix}^{-1} \cdot Park^{-1} \begin{bmatrix} \hat{i}_d \\ \hat{i}_q \\ \hat{i}_0 \end{bmatrix} = \frac{3}{2}(\hat{V}_d \hat{i}_d + \hat{V}_q \hat{i}_q + 2\hat{V}_0 \hat{i}_0), \tag{13}$$

where $\hat{V}_d$, $\hat{V}_q$, and $\hat{V}_0$ are the voltages, and $\hat{i}_d$, $\hat{i}_q$, and $\hat{i}_0$ are the currents. Furthermore, the voltages in $dq0$ coordinates can be described as

$$\begin{cases} \hat{V}_d = R_s \hat{i}_d + L_s \frac{d\hat{i}_d}{dt} - \omega_G L_s \hat{i}_q \\ \hat{V}_q = R_s \hat{i}_q + L_s \frac{d\hat{i}_q}{dt} + \omega_G L_s \hat{i}_d + \omega_G \psi_G \\ \hat{V}_0 = R_s \hat{i}_0 + L_s \frac{d\hat{i}_0}{dt} \end{cases}, \tag{14}$$

where R_s is the resistance, L_s is the inductance, ω_G is the angular frequency, and ψ_G is the magnetic linkage.

In the form of $dq0$ coordinates with a winding delta connection, the zero-sequence voltage H does not exist. Therefore, Equation (13) can be rewritten as

$$\hat{P} = \frac{3}{2}(\hat{V}_d \hat{i}_d + \hat{V}_q \hat{i}_q). \tag{15}$$

Substituting Equation (14) into (15), the active power output of a linear generator can be rewritten as

$$\hat{P} = \frac{3}{2}\left(R_s\hat{i}_d^2 + R_s\hat{i}_q^2\right) + \frac{3}{2}L_s\left(\hat{i}_d\frac{d\hat{i}_d}{dt} + \hat{i}_q\frac{d\hat{i}_q}{dt}\right) + \frac{3}{2}\omega_G\psi_G\hat{i}_q. \tag{16}$$

On the right side of Equation (16), the first term is copper loss, the second term is the increasing rate of magnetic energy, and the last term is the electromagnetic power.

According to the structure of a linear generator, the relationship between the linear generator's angular frequency ω_G and speed $\hat{v}_z$ is

$$\hat{v}_z = 2f_G\tau = 2\frac{\omega_G}{2\pi}\tau = \frac{\omega_G}{\pi}\tau \Rightarrow \omega_G = \frac{\pi}{\tau}\hat{v}_z, \tag{17}$$

where speed $\hat{v}_z$ is also the relative speed between the outer buoy and inner buoy (see the inner buoy's structure in Figure 5b), τ is the pole pitch, and f_G is the current frequency.

Substituting Equation (17) into the electromagnetic power term of Equation (16), and ignoring the first term (copper loss) and second term (increasing rate of magnetic energy), the electromagnetic power of a linear generator can be described as

$$\hat{P}_{em} = \frac{3}{2}\frac{\pi\psi_G}{\tau}\hat{i}_q\hat{v}_z. \tag{18}$$

For a linear generator, the relationship between electromagnetic power and piston speed is $\hat{P}_{em} = -\hat{F}_u\hat{v}_z$. Thus, the load force $\hat{F}_u$ can be expressed using the q-axis current $\hat{i}_q$ of $dq0$ coordinates.

$$\hat{F}_u = -\frac{3}{2}\frac{\pi\psi_G}{\tau}\hat{i}_q. \tag{19}$$

4.2. Optimized Control of Double-Buoy OWECS

According to the mechanical vibration theory [13], only when the vertical direction motion phase between the outer buoy and ocean waves is identical (in the resonance condition) can the operational efficiency of double buoys type OWECS be improved. Equations (10), (16), and (19) indicate that the vertical direction motion phase between the outer buoy and ocean waves can be synchronized by adjusting the q-axis current $\hat{i}_q$ and d-axis current $\hat{i}_d$. Therefore, on the basis of the q-axis current $\hat{i}_q$ and d-axis current $\hat{i}_d$ of a linear generator, a sliding mode control method is proposed to improve the operational efficiency of a double-buoy OWECS.

Figure 10a shows the detailed control structure of a double-buoy OWECS, including the ocean wave's vertical direction speed $\hat{v}_z^*$, outer buoy's vertical direction speed $\hat{v}_z$, sliding mode control method (SMC), proportional integral (PI) control method, linear generator, Park transformation, power inverter, and space vector pulse width modulation (SVPWM). If the detailed signal transmission process of the voltages and currents in Figure 10a is ignored, the simplified control structure of a double-buoy OWECS is shown in Figure 10b. In Figure 10b, an interference term is added to test the anti-interference ability of the sliding mode control method.

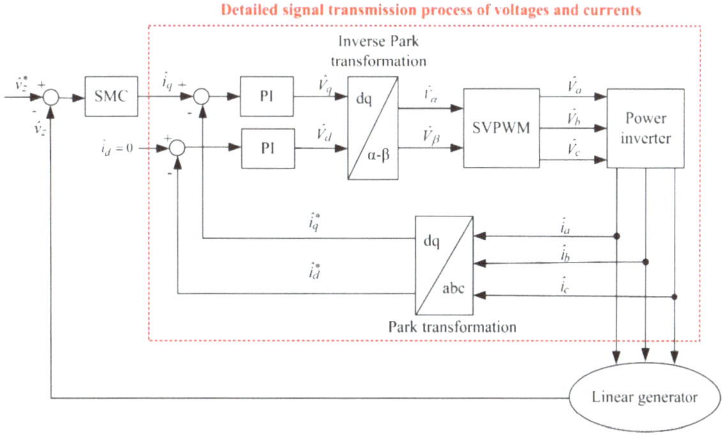

(a) Detailed control structure

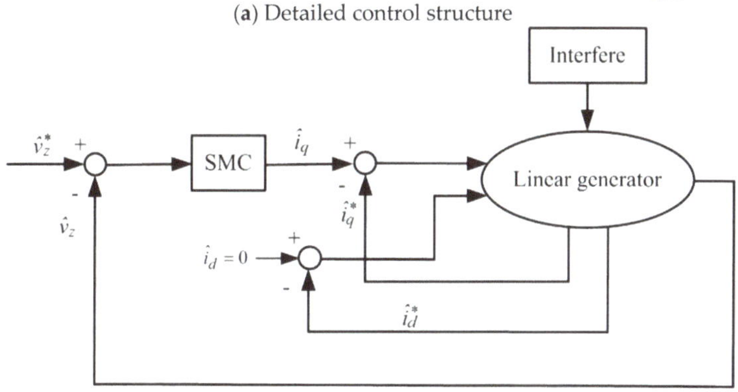

(b) Simplified control structure (block diagram)

Figure 10. The control structure of double-buoy OWECS.

According to the principle of mechanical motion [21,22], the motion equation of a linear generator can be written as

$$J\frac{d\omega_m}{dt} = T_L - T_e, \tag{20}$$

$$T_e = \frac{3}{2}P_n i_q \psi_f, \tag{21}$$

where J is the moment of inertia, ω_m is the mechanical angular speed, T_L is the driving torque, T_e is the electromagnetic torque, P_n is the pole-pair number, and ψ_f is the flux linkage.

The method of sliding mode control is adopted [23,24], and the d-axis current is set to $i_d = 0$; then, the expression of q-axis current control can be obtained as

$$i_q^* = \frac{2J}{3P_n \psi_f} \int_0^t \left(c\frac{3}{2J}P_n \psi_f i_q - g\mathrm{sgn}(s) - ks \right) dt, \tag{22}$$

where c is the coefficient of the sliding mode surface, k is the constant-velocity approach rate, $\mathrm{sgn}(s)$ is the symbolic function, and g is the exponential approach rate.

Assuming that the moment of inertia of the linear generator is $J = 2$ and the interference term is $d(t) = 10\sin(\omega_g t)$, the simulation optimization control result of the double-buoy OWECS according to the sliding mode control method is shown in Figure 11.

Figure 11a is the simulation control result with a regular ocean wave, while Figure 11b is that with an irregular ocean wave. Figure 11 indicates that the vertical direction motion phase between ocean wave and linear generator (also is the outer buoy) can be synchronized using the appropriate parameter settings of the sliding mode control method; thus, the operational efficiency of a double-buoy OWECS can be improved.

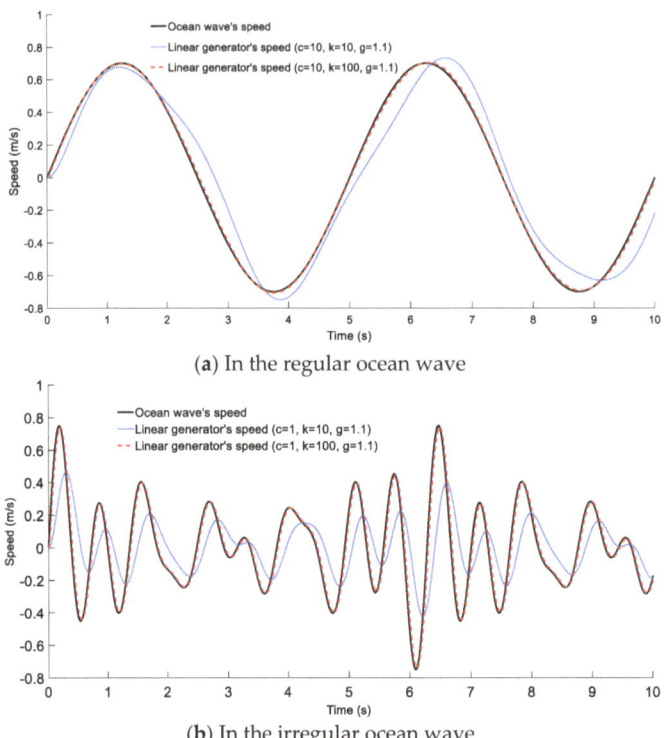

Figure 11. Optimized control of double-buoy OWECS.

5. Discussion

On the basis of modeling and experimental analysis, an optimized control method was proposed to improve the conversion efficiency of a double-buoy OWECS. However, there are several points should be considered before the practical application of optimized control of double-buoy OWECSs.

Firstly, some details of the optimized control method should be further investigated, such as the signal processing of the linear generator, hardware circuit design, and electronic component selection.

Secondly, the best match between the outer buoy and linear generator should be analyzed. For example, if the maximum output electromagnetic torque of the linear generator is less than the vertical direction ocean wave force of the outer buoy, the optimized control of a double-buoy OWECS will not be realized.

Lastly, due to the existence non-sinusoidal and irregular ocean waves, a small prototype test needs to be implemented before the ocean test, which is beneficial to the improvement and perfection of a double-buoy OWECS. In the process of small prototype tests, the structure size design, parameter setting of control method, anti-interference performance and so on should be tested and improved.

In general, the principal restrictions of study in this paper were the experimental test of the OWECS, the safety of the OWECS in a harsh ocean environment (hurricanes,

typhoons, etc.), the efficiency of the SMC method, and the ratio of investment and output, which should be considered and analyzed. For example, after about 3 days of the ocean test (see Figure 7), the OWECS was damaged by the high amplitude of ocean waves during a typhoon. Therefore, from the perspective of scientific research, the safety and high efficiency of the OWECS will be the main research topics in the near future.

6. Conclusions

In this paper, the motion model of a double-buoy OWECS was analyzed, and the correctness of the motion model was verified using an experimental test. However, the analysis of experimental test results indicated that the efficiency of the double-buoy OWECS was lower. Therefore, a sliding mode control method based on a linear generator was proposed to improve the efficiency of the double-buoy OWECS, and some simulation analysis results were presented. After modeling, experimental analysis, and optimized control of the double-buoy OWECS, a discussion was carried out, which may benefit future ocean tests of OWECSs.

Author Contributions: Conceptualization, Z.C. and X.L.; methodology, Z.C.; writing—original draft preparation, Z.C. and Y.C.; writing—review and editing, L.H. All authors have read and agreed to the published version of the manuscript.

Funding: This work was financially supported by the Scientific and Technological Project in Henan Province under Grant No. 222102240037 and No. 222102240106.

Institutional Review Board Statement: Not applicable.

Informed Consent Statement: Not applicable.

Data Availability Statement: Some or all data and models generated or used during the study are available in a repository or online.

Conflicts of Interest: The authors declare no conflict of interest.

References

1. Mustapa, M.A.; Yaakob, O.B.; Ahmed, Y.M.; Rheem, C.K.; Koh, K.K.; Adnan, F.A. Wave energy device and breakwater integration: A review. *Renew. Sustain. Energy Rev.* **2017**, *77*, 43–58. [CrossRef]
2. Brodersen, K.M.; Bywater, E.A.; Lanter, A.M.; Schennum, H.H.; Furia, K.N.; Sheth, M.K.; Kiefer, N.S.; Cafferty, B.K.; Rao, A.K.; Garcia, J.M.; et al. Direct-drive ocean wave-powered batch reverse osmosis. *Desalination* **2022**, *523*, 115393. [CrossRef]
3. Fischer, A.; Silva, J.S.; Beluco, A. Feasibility limits for a hybrid system with ocean wave and ocean current power plants in southern coast of brazil. *Comput. Water Energy Environ. Eng.* **2021**, *10*, 104581. [CrossRef]
4. Viet, N.V.; Wang, Q.; Carpinteri, A. Development of an ocean wave energy harvester with a built-in frequency conversion function. *Int. J. Energy Res.* **2018**, *42*, 684–695. [CrossRef]
5. Edwards, E.C.; Yue, K.P. Optimisation of the geometry of axisymmetric point-absorber wave energy converters. *J. Fluid Mech.* **2022**, *933*, 1–17. [CrossRef]
6. Qin, H.; Tan, S.; Xia, Z.; Zhu, Y. An analysis of international patents on ocean wave energy. *Libr. Inf. Stud.* **2012**, *4*, 45–53.
7. Cheng, Y.; Fu, L.; Dai, S.; Collu, M.; Cui, L.; Yuan, Z.; Incecik, A. Experimental and numerical analysis of a hybrid WEC-breakwater system combining an oscillating water column and an oscillating buoy. *Renew. Sustain. Energy Rev.* **2022**, *169*, 112909. [CrossRef]
8. Cheng, Y.; Fu, L.; Dai, S.; Collu, M.; Ji, C.; Yuan, Z.; Incecik, A. Experimental and numerical investigation of WEC-type floating breakwaters: A single-pontoon oscillating buoy and a dual-pontoon oscillating water column. *Coast. Eng.* **2022**, *177*, 104188. [CrossRef]
9. Jia, C.; Cao, H.; Pan, H.; Ahmed, A.; Jiang, Z.; Azam, A.; Zhang, Z.; Pan, Y. A wave energy converter based on a zero-pressure-angle mechanism for self-powered applications in near-zero energy sea crossing bridges. *Smart Mater. Struct.* **2022**, *31*, 095006. [CrossRef]
10. Baghbani Kordmahale, S.; Do, J.; Chang, K.A.; Kameoka, J. A hybrid structure of piezoelectric fibers and soft materials as a smart floatable open-water wave energy converter. *Micromachines* **2021**, *12*, 1269. [CrossRef] [PubMed]
11. Zou, S.; Abdelkhalik, O. Modeling of a variable-geometry wave energy converter. *IEEE J. Ocean. Eng.* **2021**, *46*, 879–890. [CrossRef]
12. Park, J.S.; Gu, B.G.; Kim, J.R.; Cho, I.H.; Jeong, I.; Lee, J. Active phase control for maximum power point tracking of linear wave generator. *IEEE Trans. Power Electron.* **2017**, *32*, 7651–7662. [CrossRef]
13. Falnes, J. *Ocean Waves and Oscillating Systems*; Cambridge University Press: Cambridge, MA, USA, 2002.
14. Farrok, O.; Islam, M.R.; Sheikh, M.R.; Guo, Y.; Zhu, J.; Xu, W. A novel superconducting magnet excited linear generator for wave energy conversion system. *IEEE Trans. Appl. Supercond.* **2016**, *26*, 1–5. [CrossRef]

15. García-Tabarés, L.; Lafoz, M.; Blanco, M.; Torres, J.; Obradors, D.; Nájera, J.; Navarro, G.; García, F.; Sánchez, A. New type of linear switched reluctance generator for wave energy applications. *IEEE Trans. Appl. Supercond.* **2020**, *30*, 19642959.
16. Xia, T.; Yu, H.; Guo, R.; Liu, X. Research on the field-modulated tubular linear generator with quasi-halbach magnetization for ocean wave energy conversion. *IEEE Trans. Appl. Supercond.* **2018**, *28*, 17610493. [CrossRef]
17. Viet, N.V.; Xie, X.D.; Liew, K.M.; Banthia, N.; Wang, Q. Energy harvesting from ocean waves by a floating energy harvester. *Energy* **2016**, *112*, 1219–1226. [CrossRef]
18. Masoumi, M.; Wang, Y. Repulsive magnetic levitation-based ocean wave energy harvester with variable resonance: Modeling, simulation and experiment. *J. Sound Vib.* **2016**, *381*, 192–205. [CrossRef]
19. Xie, D.M.; Chen, Y.P.; Zhang, C.K. On wave distribution of the East China Sea. *Port Waterw. Eng.* **2012**, *11*, 14–21.
20. Masamichi, I.; Shinji, D. PMSM Model Discretization in Consideration of Park Transformation for Current Control System. In Proceedings of the International Power Electronics Conference, Niigata, Japan, 20–24 May 2018.
21. Liu, K.; Hou, C.; Hua, W. A Novel Inertia Identification Method and Its Application in PI Controllers of PMSM Drives. *IEEE Access* **2016**, *7*, 13445–13454. [CrossRef]
22. Wang, P.; Xu, Y.; Ding, R.; Liu, W.; Shu, S.; Yang, X. Multi-Kernel Neural Network Sliding Mode Control for Permanent Magnet Linear Synchronous Motors. *IEEE Access* **2021**, *9*, 57385–57392. [CrossRef]
23. Liu, J. *Sliding Mode Control Design and MATLAB Simulation: The Basic Theory and design Method*; Tsinghua University Press: Beijing, China, 2015.
24. Wei, Y.; Sun, L.; Chen, Z. An improved sliding mode control method to increase the speed stability of permanent magnet synchronous motors. *Energies* **2022**, *15*, 15176313. [CrossRef]

Article

Numerical Study on a Cylinder Vibrator in the Hydrodynamics of a Wind–Wave Combined Power Generation System under Different Mass Ratios

Hongyuan Sun [1], Jiazheng Wang [1], Haihua Lin [1,*], Guanghua He [2,*], Zhigang Zhang [2], Bo Gao [1] and Bo Jiao [1]

1 College of Naval Architecture and Port Engineering, Shandong Jiaotong University, Weihai 264209, China
2 School of Ocean Engineering, Harbin Institute of Technology, Weihai 264209, China
* Correspondence: linhaihua@sdjtu.edu.cn (H.L.); gh.he@hit.edu.cn (G.H.)

Abstract: A hydrodynamic wind–wave combined power generation system is a new type of energy device that uses wind and ocean current energy to generate electricity. In this paper, the hydrodynamics of a wind–wave combined power generation system was simulated in Fluent. The fluid–structure coupling simulation of the vortex vibration of the cylindrical oscillator was realized using UDF and dynamic mesh technology. The Vortex-Induced Vibration (VIV) characteristics of the cylindrical oscillator were analyzed, and the reliability of the numerical simulation method was verified by comparing the amplitude and trajectory of the eddy-excited vibration with the classic experiments of Jauvtis and Williamson. The VIV characteristics of cylindrical oscillators with different mass ratios were studied in terms of vibration response, motion trajectory, and the streamwise equilibrium position. The effect of the mass ratio on the hydrodynamics of a wind–wave combined power generation system was simulated using spring damping, achieving the goal of carrying out preliminary research work simulating the wind–wave combined power generation device. Some useful conclusions were obtained through calculation, which provided data support for the corresponding platform device. This study shows that in cylindrical oscillators with different mass ratios, the overall trend at the same reduced velocity is that the larger the mass ratio, the smaller the crossflow amplitude. The cylindrical oscillators with mass ratios of one and two appear in the upper branch, while cylindrical oscillators with mass ratios of three and four do not appear, and with the increase in the mass ratio, the frequency ratio in the lower branch tends toward one. At the same reduced velocity, the lower the mass ratio, the larger the corresponding downstream equilibrium position, and the higher the energy acquisition efficiency.

Keywords: wind–wave combined; current energy; vortex-induced vibration; cylindrical oscillator; mass ratio; numerical simulation

Citation: Sun, H.; Wang, J.; Lin, H.; He, G.; Zhang, Z.; Gao, B.; Jiao, B. Numerical Study on a Cylinder Vibrator in the Hydrodynamics of a Wind–Wave Combined Power Generation System under Different Mass Ratios. *Energies* **2022**, *15*, 9265. https://doi.org/10.3390/en15249265

Academic Editor: Muhammad Aziz

Received: 10 November 2022
Accepted: 4 December 2022
Published: 7 December 2022

Publisher's Note: MDPI stays neutral with regard to jurisdictional claims in published maps and institutional affiliations.

Copyright: © 2022 by the authors. Licensee MDPI, Basel, Switzerland. This article is an open access article distributed under the terms and conditions of the Creative Commons Attribution (CC BY) license (https://creativecommons.org/licenses/by/4.0/).

1. Introduction

Marine renewable energy sources mainly include offshore wind, wave, and tidal energy. Among them, wind energy is generally regarded as the best alternative to fossil energy due to its abundant resources and the fact that it is renewable and does not produce greenhouse gases. Offshore wind energy resources are extremely abundant, and the large-scale development of offshore wind power is an important means to prevent and control air pollution. In addition, as a kind of renewable green energy, ocean current energy has the characteristics of high sustainability, high energy density, and abundant reserves, it has broad development prospects [1,2].

At present, the main types of equipment for the development of ocean current energy are the parachute type and magnetic flow type. However, this type of energy generation requires a high velocity, so it is difficult to apply it in waters with large water depths and low velocity. The use of vortex-induced vibration for power generation, which does not require high velocity and low cost, has become a research hotspot in the field of sea current

energy generation in recent years [3]. At the same time, with the continuous construction of land-based wind farms, the limited development of land-based wind energy resources has become increasingly saturated. Offshore wind power generation is gradually being developed because of its unique advantages. Therefore, can the combination of the two power-generation devices produce more electricity for people to use?

Through this research, it was found that the wind–wave combined power generation system can take advantage of the volatility, intermittency, irregularity, and stability of offshore wind power, and wave energy is more stable compared to wind power [4]. Therefore, the combined power generation system of the two types of energy can reduce the number of hours it does not work, compared with a wind power system alone. The VIV power generation device can also absorb the wave energy near the offshore wind turbine platform, changing the local wave field, which can effectively protect the wind turbine from strong wave impacts under reasonable arrangement, reducing the fatigue of the wind power generation device, and increasing its service life.

The overall structure of a Spar-type floating wind turbine is a deep-draft slender cylinder, which mainly consists of two parts: a floating chamber and a ballast chamber [5]. The stability of this device is maintained by uniformly distributed mooring cables connected to the seabed, so this structure not only provides the wind turbine as a whole with a large recovery moment and high inertia resistance in both the transverse and longitudinal rocking directions, but also reduces the vertical swing motion, which greatly improves the floating wind turbine in terms of water anti-tilting stability. The research in this paper is mainly based on the wind–wave combined power generation system with a Spar-type wind turbine as the main structure. By installing a VIV power generation device on this device, both wind and wave energy can be utilized, and the cost can be reduced to improve the overall economic efficiency.

Based on the background described above, the first study was carried out to investigate the parametric influence of the VIV power generation device on the wind–wave combined power generation system (mainly the influence of the mass ratio on the energy gain efficiency of the device) as a prelude to the future study of the power generation efficiency of the wind–wave combined power generation system.

Cylindrical vortex-induced vibration is a complex fluid–solid coupling phenomenon involving fluid mechanics, vibration mechanics, structural mechanics, computational fluid mechanics, and other disciplines [6,7]. A vortex-induced vibration generator is an energy conversion device that captures the energy of sea currents by using the vortex vibration effect. When the water flows over the body vibration, under certain conditions, it will cause a body for VIV; this vibration can transform the mechanical energy of the current into the kinetic energy of the vibrating body, followed by the kinetic energy of the vibrator to the generator (generator stator and rotor). The conversion of kinetic energy of the vibrator into mechanical energy will then move the sub, cut the magnetic induction line, generate electricity, and then store the electricity.

The concept of Vortex-Induced Vibration for Aquatic Clean Energy (VIVACE) was first proposed by Professor Bernitsas' team at the University of Michigan [8], and the experimental model of VIVACE was also developed by his team [9], and mainly consists of three parts: a cylindrical oscillator, a transmission, and a power generator.

Power generation can be divided into two types according to the different types of energy acquisition: piezoelectric and electromagnetic. The method of converting the mechanical energy of the cylindrical oscillator into electrical energy by the positive voltage effect of the piezoelectric polymer is called piezoelectric, and the representative devices include vertical moving cylindrical piezoelectric energy-harvesting devices and piezoelectric bluff body energy-harvesting devices. The method of converting the kinetic energy generated by the vibration of the vibrator into drive energy and driving the generator rotor to cut the magnetic field lines for energy conversion is called the electromagnetic type, and the representative device is the VIVACE device [10]. At present, the research on the piezoelectric VIV power generation device is still in the initial stages, mainly because

of its low energy-acquisition efficiency, small energy-acquisition flow range, and poor energy-acquisition stability; meanwhile, the electromagnetic power generation device has greater energy acquisition, higher energy acquisition efficiency, and wider synchronous vibration range. Therefore, the electromagnetic power generation device is more suitable for practical engineering applications [8].

Many studies have been conducted to date on the parameterization of VIV generators, and some of them have been relatively well studied. There have also been some reviews discussing studies on the parameterization of vortex-induced vibration (e.g., Lin 2015; Huang 2007). Huang et al.'s study showed that when the mass ratio is greater than 3.5, there is little effect on the transverse amplitude, regardless of whether the flow direction motion is restricted, while the motion of two-way degrees of freedom can produce a larger transverse amplitude when the mass ratio is less than 3.5 [11]. Du et al. found that the mass ratio has a significant effect on the VIV of a double cylindrical structure connected in series, both with two degrees of freedom [12]. Bai and Chen found that with the gradual increase in the diameter ratio, the amplitude ratio of the columns of the series' unequal-diameter cylindrical power generation system changes with the size of the columns; at the same time, the existence of a diameter ratio less than 0.7 makes the amplitude ratio of the large and small columns achieve the maximum value; when the spacing is relatively small, the change in the diameter ratio has a greater impact on the amplitude response [13]. Tan et al. found that under the analysis of energy utilization, in the upper part of the branch, the product of cylindrical oscillator amplitude and vibration frequency is the largest in the uniform flow motion, and its effective power is higher in uniform flow than in turbulent flow [14]. Khalak and Williamson found that under low mass and damping conditions, three corresponding branches exist, namely the initial branch (the 2S mode: two independent vortices are released per cycle), the upper branch (2P mode: a pair of vortices are released every half cycle), and the lower branch (also with 2P mode) [15]. Yang et al. found that with the increase in the mass ratio, the maximum amplitude in the transverse direction decreases at a small, reduced velocity. With the decrease in the mass ratio, the influence of the phase difference between the lift and the displacement on the amplitude of the downstream cylinder is more significant [16]. In addition, Han found that the downstream amplitudes of both rough and smooth cylindrical oscillators showed a trend of increasing and then decreasing. However, with the increase in the flow velocity, the smooth cylindrical oscillator no longer produces vortex resonance, and the amplitude and then the energy gain efficiency decrease rapidly; meanwhile, with the rough oscillator lift and motion direction, the vortex vibration moves into the galloping stage, and the amplitude is constantly rising [17]. Continuing research on the effect of the mass ratio on VIV can enrich this part, which is important for better promotion of clean and sustainable ocean energy in the future, and also has a deeper and broader significance for the development of new forms of power-generation devices.

The difference between this paper and previous work is that a parametric study was used for fluid–solid coupling and wind–wave combined power generation in this paper. In this study, the effect of the mass ratio on the hydrodynamics of the wind–wave combined power generation system of cylindrical oscillators was focused and analyzed using numerical simulations with the VIV energy acquisition efficiency formula. The reliability of the numerical simulation method is demonstrated by comparing it with the classic experiments of Jauvtis and Williamson in terms of response amplitude and motion trajectory, and the characteristics of cylindrical oscillators with different mass ratios in terms of VIV response, motion trajectory, motion frequency, and downstream equilibrium position are investigated.

2. Non-Dimensional Groups and Energy Acquisition Efficiency

Figure 1 shows the wind–wave combined power generation system with a Spar-type wind turbine as the main structure.

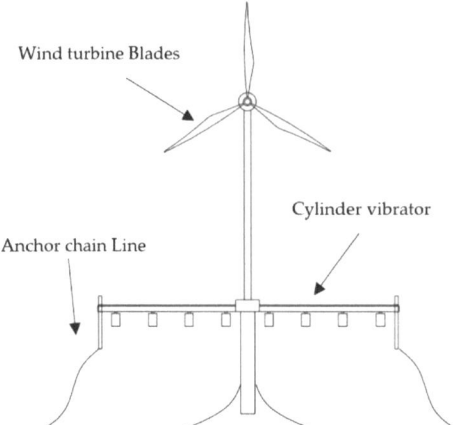

Figure 1. The Hydrodynamics of a Wind–Wave Combined Power Generation System.

The relevant parameters involved are as follows:

$$Reynolds\ number: Re = \frac{\rho U D}{\mu} = \frac{U D}{\nu} \qquad (1)$$

In this formula, U is the incoming flow velocity; D is the diameter of the cylindrical oscillator; ρ is the density of the fluid; μ is the fluid dynamic coefficient; and ν is the fluid kinematic viscosity coefficient.

$$Strouhal\ number: St = \frac{f_s D}{U} \qquad (2)$$

In this formula, f_s is the vortex shedding frequency; U is the incoming flow velocity; and D is the diameter of the cylindrical oscillator.

$$m^*: m^* = \frac{m}{\frac{1}{4}\rho\pi D^2} \qquad (3)$$

In this formula, m is the mass of the cylindrical oscillator per unit length; ρ is the density of the fluid; and D is the diameter of the cylindrical oscillator. The m^* is the ratio of the mass of the cylindrical oscillator to the mass of the fluid displaced.

$$Reduced\ Velocity: U_r = \frac{U}{f_n D} \qquad (4)$$

In this formula, U is the incoming flow velocity; f_n is the natural frequency of the vibration system; and D is the diameter of the cylindrical oscillator. The U_r is the ratio of the length of water flowing through a self-oscillation cycle of the structure in the water to the characteristic size of the cylindrical oscillator.

$$A^*: A^* = \frac{A}{D} = \frac{A_{max} - A_{min}}{2D} \qquad (5)$$

In the formula: A is the amplitude; D is the diameter of the cylindrical oscillator; A_{max} is the maximum amplitude, and A_{min} is the minimum amplitude. Among them, the A^* is the ratio of the amplitude of the cylindrical oscillator to the diameter of the cylindrical oscillator.

The energy conversion efficiency η is dimensionless, which is defined as the ratio of captured energy to the fluid energy [13].

$$\eta = \frac{W_{FLM}}{W_{fluid}} = \frac{\frac{1}{2}C_{system}\omega_{OSC}{}^2 T_{OSC}\left(A_y{}^2+4A_x{}^2\right)}{\frac{1}{2}\rho DU^3 T_{OSC}}$$
$$= \frac{C_{system}\omega_{OSC}{}^2 T_{OSC}\left(A_y{}^2+4A_x{}^2\right)}{\rho DU^3 T_{OSC}} \quad (6)$$

In this formula, W_{FLM} is the energy of a VIV cylinder in a unit period; W_{fluid} is a uniform flow of energy over a period; C_{system} is the system damping coefficient; ω_{OSC} is the angular frequency; and A_x and A_y are the amplitudes in the corresponding direction of the cylinder.

Through the derivation of the equation, the energy acquisition efficiency formula of VIV can be finally obtained as follows:

$$\eta = \frac{4\pi^4(m^* + C_a)S_t{}^2(4x^2 + y^2)\tilde{\zeta}_{system}}{U_r} \quad (7)$$

In this formula, C_a is the fluid-added mass coefficient of the cylinder, and $\tilde{\zeta}_{system}$ is the system damping ratio coefficient.

Through analysis, we can show that the main parameters affecting the energy conversion efficiency are the mass ratio, Strouhal number, dimensionless displacements x and y, and the damping ratio coefficient, which are all proportional to the energy conversion efficiency [14,18]. Meanwhile, the energy conversion efficiency is inversely proportional to the deceleration rate. To address the issue of parameters affecting the energy conversion rate, this paper mainly compares the effects of different mass ratios on VIV.

3. Numerical Model of Vortex-Induced Vibration

Fluent software and UDF programming are used to simulate the flow field around the cylindrical oscillator, simplify the experimental model of the VIV generator, and combine the dynamic mesh technology to realize the fluid–structure coupling of the cylindrical oscillator of the vortex-induced vibration power generation device.

In this paper, the experimental model of the vortex-induced vibration generator is simplified into the following physical model, as shown in Figure 2.

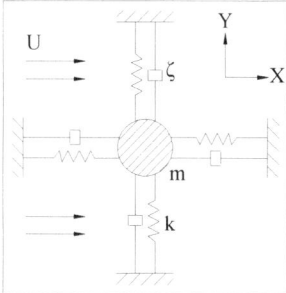

Figure 2. Vortex-induced vibration model of cylindrical vibrator.

The SST k-ω model used in the numerical simulations in this paper belongs to the Reynolds averaging method. The SST k-ω model has good results in both near-wall and far-field calculations. Compared with the standard k-ω model, the SST k-ω model takes into account the shear stress transport mode, changes the turbulence constant in the model, and performs better in solving the negative pressure gradient in turbulent flow problems. The above features give the SST k-ω model a wider range of applications.

The SST k-ω equation is:

$$\frac{\partial}{\partial t}(\rho k) + \frac{\partial}{\partial x_i}(\rho k u_i) = \frac{\partial}{\partial x_j}\left(\Gamma_k \frac{\partial k}{\partial x_j}\right) + \tilde{G}_k - Y_k + S_k \quad (8)$$

$$\frac{\partial}{\partial t}(\rho \omega) + \frac{\partial}{\partial x_i}(\rho \omega u_i) = \frac{\partial}{\partial x_j}\left(\Gamma_\omega \frac{\partial \omega}{\partial x_j}\right) + G_\omega - D_\omega + S_\omega \qquad (9)$$

In these equations, Γ_k and Γ_ω are the effective diffusion terms of k and ω; $\tilde{G}_k$ is the turbulent kinetic energy generation terms of k; G_ω is the generic term of k; Y_k is the dissipation terms of k; Y_ω is the dissipation term of ω; D_ω is the cross-diffusion term; and S_k and S_ω are the custom source terms.

After the mass force is omitted, the N-S equation can be simplified as follows:

$$\left.\begin{array}{l}\frac{\partial u}{\partial t} + \frac{\partial(uu)}{\partial x} + \frac{\partial(uv)}{\partial y} + \frac{\partial(uw)}{\partial z} = -\frac{1}{\rho}\frac{\partial p}{\partial x} + \frac{1}{\rho}\left(\frac{\partial \sigma_{xx}}{\partial x} + \frac{\partial \tau_{yx}}{\partial y} + \frac{\partial \tau_{zx}}{\partial z}\right) \\ \frac{\partial v}{\partial t} + \frac{\partial(uv)}{\partial x} + \frac{\partial(vv)}{\partial y} + \frac{\partial(vw)}{\partial z} = -\frac{1}{\rho}\frac{\partial p}{\partial y} + \frac{1}{\rho}\left(\frac{\partial \tau_{xy}}{\partial x} + \frac{\partial \sigma_{yy}}{\partial y} + \frac{\partial \tau_{zy}}{\partial z}\right) \\ \frac{\partial w}{\partial t} + \frac{\partial(uw)}{\partial x} + \frac{\partial(vw)}{\partial y} + \frac{\partial(ww)}{\partial z} = -\frac{1}{\rho}\frac{\partial p}{\partial z} + \frac{1}{\rho}\left(\frac{\partial \tau_{xz}}{\partial x} + \frac{\partial \tau_{yz}}{\partial y} + \frac{\partial \sigma_{zz}}{\partial z}\right)\end{array}\right\} \qquad (10)$$

In these formulas, u, v, w, and p are the instantaneous values of velocity and pressure, respectively.

In uniform flow, the governing equation of cylindrical vortex-induced motion is

$$m\frac{d^2 x}{dt^2} + C_x \frac{dx}{dt} + K_x x = F_d(t) \qquad (11)$$

$$m\frac{d^2 y}{dt^2} + C_y \frac{dy}{dt} + K_y y = F_l(t) \qquad (12)$$

In these formulas, t is the time; m is the mass; $C_x = 4\pi m \zeta_x f_{nx}$ and $C_y = 4\pi m \zeta_y f_{ny}$ are the damping coefficients of the forward and crossflow motions, respectively, where ζ is the damping ratio; K_x and K_y are the system stiffnesses in the forward and crossflow directions, respectively; x and y are the forward-flow directions, and crossflow VIV displacement, respectively; $F_d(t)$ is resistance; and $F_l(t)$ is lift.

In Jauvtis and Williamson's experiment, the cylindrical oscillator has a mass ratio of 2.6, the intrinsic frequency $f_{nx} = f_{ny} = 0.4$ Hz, and the reduced velocity ranges from 2.2 to 13.5 [8]. Therefore, the working conditions shown in Table 1 were chosen.

Table 1. Data table of calculation conditions.

m^*	f_{ny}/H	f_{nx}/Hz	U_C/m·s^{-1}	U_r	R_e
1	0.38	0.38	0.05~0.81	1.32~21.32	5000~81,000
2	0.38	0.38	0.05~0.69	1.32~18.16	5000~69,000
3	0.38	0.38	0.05~0.81	1.32~21.32	5000~81,000
4	0.38	0.38	0.05~0.81	1.32~21.32	5000~81,000
2.6	0.38	0.38	0.05~0.67	1.32~17.63	5000~67,000

4. Computational Domain and Numerical Simulation Reliability Verification

4.1. Computational Domain

In this paper, we ignore the subsidiary structures of the vortex-excited vibration power generator in the numerical simulation and only investigate the cylindrical oscillator.

Figures 3 and 4, for the two-dimensional cylindrical oscillator vortex-induced vibration of the schematic diagram and the flow field grid computing area, respectively, show a cylindrical stator $D = 0.1$ m diameter, an area calculation of $20\ D \times 40\ D$, a speed entrance D from the center position for 10 D from the center of the circle pressure exits for 30 D, the upper and lower boundary sliding types from the center of the circle to 10 D, and a cylinder for the non-sliding mode surface boundary. To consider the application of a dynamic grid in the numerical simulation of vortex-induced vibration, a random grid with a diameter of 7D is set around the cylindrical oscillator. The random grid is structured, and the other areas are unstructured. In the near-wall treatment, ten layers of boundary layers are set, and the mesh height of the first layer corresponds to one, as shown in Figure 4b. Since the

dynamic grid technology is used in the numerical simulation, grid-independent verification is needed. In this paper, the vortex excited motion of a cylinder with a reduced velocity is 5, which was tested under the condition of $m^* = 2.6$ and two-way natural frequency $f_{nx} = f_{ny} = 0.38$ Hz. Grid-irrelevance verification results are shown in Table 2. Grid B is selected as the optimal scheme through comparison.

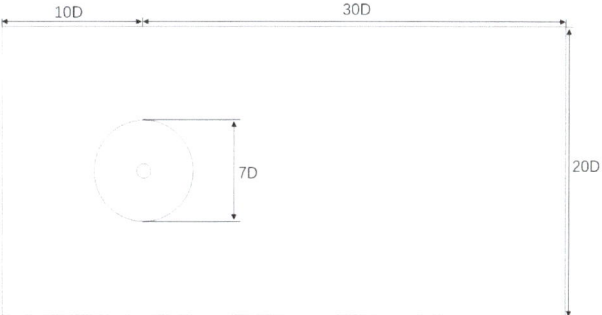

Figure 3. Calculation area.

(a) Overall mesh generation of the flow field

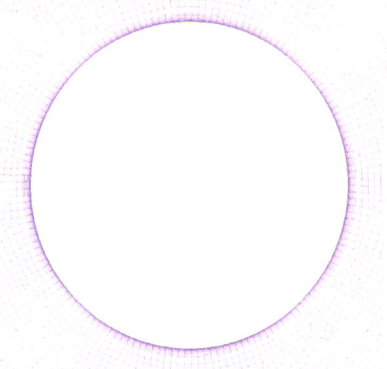

(b) Cylindrical vibrator near flow field network

Figure 4. Flow field mesh generation.

4.2. Numerical Simulation Reliability Verification

In this paper, we ignore the subsidiary structures of the vortex-excited vibration power generator in the numerical simulation and only investigate the cylindrical oscillator.

Table 2. Grid-Irrelevance Verification Results.

Grid	A	B	C
Number of units	15,328	25,554	35,452
A^*_{ymax}	0.890	0.979	0.979
A^*_{xmax}	0.155	0.176	0.178
Cl_{max}	3.124	3.697	3.698
Cd_{max}	2.013	2.174	2.176

In order to verify the reliability of the numerical simulation, the cylindrical oscillator with a mass ratio of 2.6 in Table 1 was selected and compared with the classical experiment of Jauvtis and Williamson from 2004 [9] in terms of vortex-induced vibration amplitude and trajectory.

Figure 5 compares the dimensionless amplitudes with the experimental results at different reduction speeds. As can be seen from Figure 5a, the variation trend of the A^*_y curve of dimensionless transverse flow amplitude, obtained by numerical simulation, is the same as that obtained by experiment, showing the initial branch (I), upper branch (U), lower branch (L), and desynchronization (D).

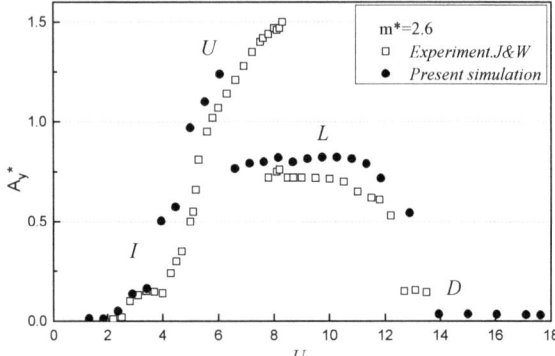

(a) Amplitude of vortex-induced vibration in the transverse direction

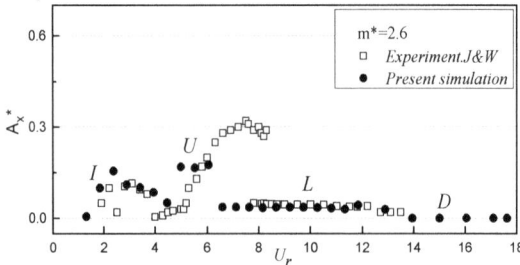

(b) Amplitude of vortex-induced vibration in the downstream direction

Figure 5. Comparison of non-dimensional amplitude with experimental results under different reduced velocities.

As shown in Figure 5b, the trend of the dimensionless forward flow amplitude A^*_x curve obtained by the numerical simulation is basically consistent with that obtained by the experiment, except that the forward-flow amplitude of the upper branch is somewhat underestimated. The reasons for this situation are as follows: in the upper branch, the vibration amplitude of the cylindrical oscillator is relatively large, which attenuates the axial correlation of the cylindrical oscillator. In the numerical simulation of two-dimensional

cylindrical oscillators, it is assumed that their axial directions are completely correlated [11]. In addition, the numerical simulation uses the Reynolds averaging method, which does not account for the random disturbances in the flow field [19,20]. By comparison, it can be concluded that the application of numerical calculations can better simulate the VIV of cylindrical oscillators, but there is an underestimation in terms of the maximum amplitude peak.

Figure 6 shows a comparison of the trajectory of the cylindrical oscillator at different reduced velocities between the numerical simulation and the trajectory obtained from Jauvtis and Williamson's experiment. The trajectories of the cylindrical oscillator obtained from the numerical simulation for the initial branch, the upper branch, and the lower branch are in good agreement with the experimental results, which again proves the reliability of the numerical simulation method.

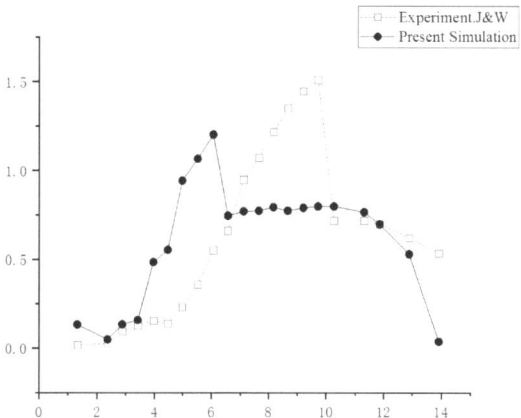

Figure 6. The trajectory of the cylindrical oscillator at different reduced velocities.

5. Effect of Mass Ratio on Vortex-Induced Vibration of Cylindrical Oscillators

According to the working conditions listed in Table 2, the vortex-induced vibration characteristics of the cylindrical oscillator of the vortex vibration power generation device with mass ratios of one, two, three, and four are studied by numerical simulation methods. Additionally, the influence of the change in mass ratio on the vortex vibration characteristics of the cylindrical oscillator is analyzed in terms of vortex vibration response, motion trajectory, motion frequency, and downstream equilibrium position.

5.1. Vortex-Induced Vibration Response

Figure 7 gives the variation law of the dimensionless vortex vibration amplitude with the reduced velocity for cylindrical oscillators with different mass ratios. Compared with the four different mass ratios of cylindrical oscillators, the crossflow vortex-induced vibration amplitude shows a trend of increasing and then decreasing with the increase in the reduced velocity, while the crossflow amplitude responses of cylindrical oscillators all show the initial branch and the lower branch.

When the mass ratios are one and two, the transverse amplitude also appears in the super-upper branch, the reduced velocities of the super-upper branch and the lower branch dividing point are both greater than six, and the transverse amplitudes are both greater than 1.25 D. When the mass ratios are three and four, the reduced velocities of the initial branch's and the lower branch's dividing point are five. In general, with the same reduced velocities, the larger the mass ratio of the cylindrical oscillator, the smaller the transverse amplitude, which is especially obvious in the initial branch and the first half of the lower branch [20,21].

In the upper branch, the transverse motion amplitude of the cylindrical oscillator with a mass ratio of two is larger than that of the cylindrical oscillator with a mass ratio of one for the same reduced velocity ($U_r = 6.5$). The reason for this phenomenon is that at this stage, the amplitude of the downstream motion for a mass ratio of one is significantly larger than that of the downstream motion for a mass ratio of two. It can be said that in the upper branch, the downstream vortex motion of the cylindrical oscillator "consumes" the energy of the cross-stream motion.

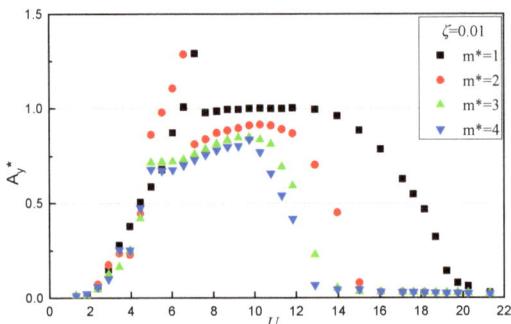

(a) Amplitude of vortex-induced vibration in the transverse direction

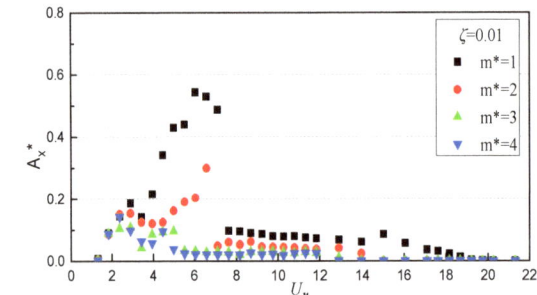

(b) Amplitude of vortex-induced vibration in the streamwise direction

Figure 7. The variation in the amplitude of the dimensionless vortex-induced motion of the cylindrical oscillator under different mass ratios with Ur.

When the mass ratio is less than two, the amplitude of vortex-induced vibration in the streamwise direction also shows the above trend. When the mass ratio is greater than or equal to three, the variation in the lower branch of the amplitude in the streamwise direction is relatively stable and not affected by the reduction speed. When the mass ratio is one and two and the reduced velocity is between six and eight, the amplitude of the downstream direction corresponding to the transverse amplitude of the lower branch has a peak value; that is, the amplitude of the downstream direction increases, and the amplitude of the transverse direction decreases rapidly. With the increase in the reduced velocity, the down-flow and its amplitude decrease.

The vortex-induced vibration downstream amplitude also shows the above trend in the case of a mass ratio less than or equal to two, while the change in the downstream amplitude of the lower branch of the mass ratio greater than or equal to three is relatively smooth and is not greatly affected by the reduced velocity. When the mass ratio is one and two and the reduced velocity is between six and eight, the downstream amplitude corresponding to the lateral amplitude of the lower branch shows a peak, during which the downstream amplitude increases and the cross-stream amplitude decreases rapidly [22]; with the increase in the reduced velocity, both the downstream amplitude and the cross-stream amplitude decrease continuously.

The above analysis shows that a cylindrical oscillator with a low mass ratio will produce a larger crossflow amplitude than a cylindrical oscillator with a high mass ratio at the same reduced velocity, which can improve the power generation efficiency of the hydrodynamics of the wind–wave combined power generation system device to a certain extent.

5.2. Effect of Different Mass Ratios on Capacitation Efficiency

According to the energy-acquisition efficiency equation [13], a single cylindrical oscillator structure is used to ensure that the intrinsic frequency and diameter remain unchanged, and the mass ratio is changed by adjusting the length-to-diameter ratio to analyze the effect of the mass ratio on the energy acquisition efficiency. The cylindrical oscillator is simulated numerically, and the two-way coupling calculation is carried out at different flow rates (i.e., different deceleration velocities) to obtain the stable amplitude of the cylindrical vortex-induced vibration for dimensionless processing. The results are shown in Figure 8.

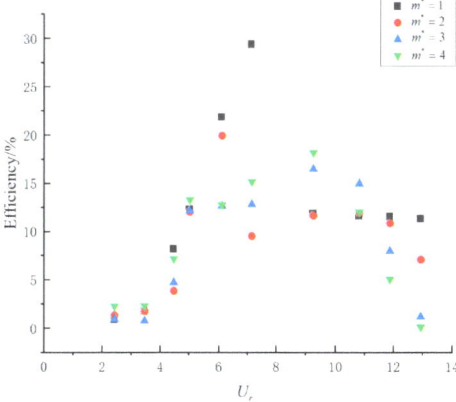

Figure 8. The relationship between mass ratio and energy efficiency.

Figure 8 shows that the energy-acquisition efficiency of cylindrical oscillators with mass ratios of one and two increases with the increase in the reduced velocity, when the reduced velocity is less than seven, and it decreases otherwise. For cylindrical oscillators with mass ratios of three and four, the maximum energy acquisition efficiency appears in the range of reduced velocity from eight to ten. In general, the energy acquisition efficiency of cylindrical oscillators increases with the increase in reduced velocity and then decreases gradually. Under the condition of the same reduction speed, the smaller the mass ratio, the higher the capacitive efficiency.

5.3. Movement Trajectory

Table 3 shows the trajectories of cylindrical oscillators with mass ratios of one, two, three, and four at different reduced velocities; the horizontal coordinates in the figure are the causeless streamwise displacements, and the vertical coordinates are the causeless transverse displacements of the cylinder [21,23]. With the increase in the reduced velocity, the amplitude also starts to increase, and the periodicity of the downstream and transverse displacement time history curves is enhanced. When the mass ratio is two and the reduced velocity is 6.05, the frequency of the downstream vortex force generated by the vortex-induced vibration of the cylindrical oscillator without restricting the flow direction is twice the frequency of the transverse vibration [24], and a more classical "8"-shaped motion trajectory appears.

When the reduced velocity is small, the vortex-induced vibration of the cylindrical oscillator is in the initial branch, the transverse motion amplitude is small, the periodicity of

the time history curve between the forward and lateral displacements is poor, the forward motion is dominant, and the "8"-shaped motion trajectory is not obvious, as shown in Table 2 when the reduced velocity is 2.37.

Table 3. Motion trajectory of the cylindrical oscillator at different reduced velocities.

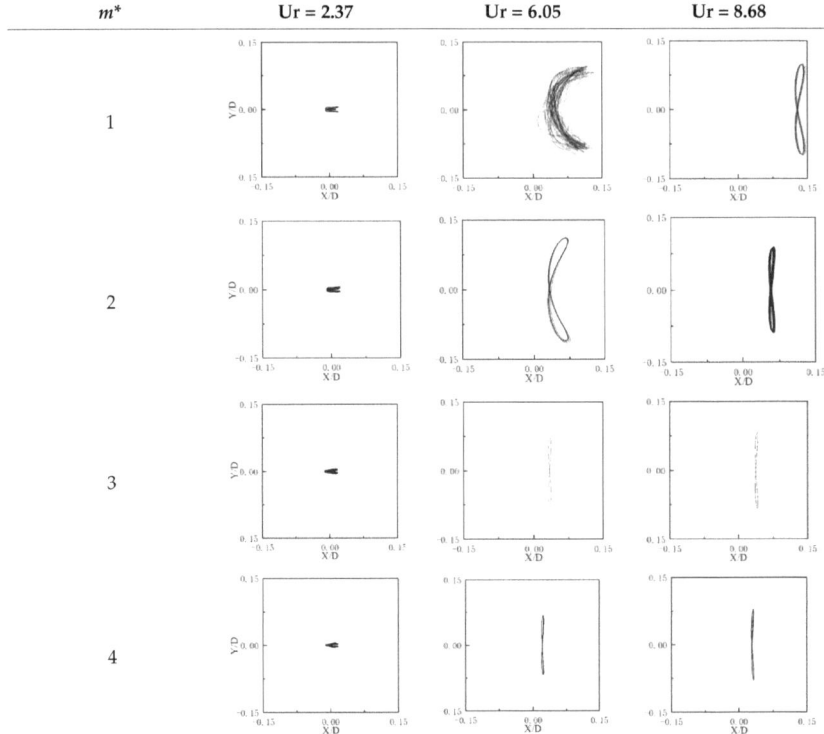

When the reduced velocity is small, the vortex-induced vibration of the cylindrical oscillator is in the initial branch, the amplitude of the transverse motion is small, the periodicity of the downstream displacement and the transverse displacement is poor, the downstream motion is dominant, and the "8"-shaped motion trajectory is not obvious, such as the motion trajectory when the reduced velocity is 2.37 [25]. This is shown in Table 3, in which X/D is the dimensionless displacement in the downstream direction and Y/D is the dimensionless displacement in the transverse direction.

As can be seen from Table 3, with the increase in the reduced velocity, the displacement in the down-flow and crossflow directions of the remaining three conditions, except for the mass ratio of one, decreases rapidly, showing a long and thin "8" shape. Furthermore, in the initial branch, with the reduced velocity of 2.37, the trajectory of the cylinder is not affected much by the Strouhal number and drag force; in the upper branch, with the reduced velocity of 6.05, the trajectory of the cylinder is more affected by the Strouhal number and drag force; in the lower branch, with the reduced velocity of 8.68, the trajectory of the cylinder is affected by the Strouhal number and drag force in approximately the same way as the upper branch.

5.4. Movement Frequency

Figure 9 shows the frequency ratio (f_y/f_n) of the crossflow motion frequency (f_y) to the structure's hydrostatic intrinsic frequency (f_n) as a function of the reduced velocity.

For comparison, the ranges of the reduced velocities corresponding to different response branches are marked in the figure.

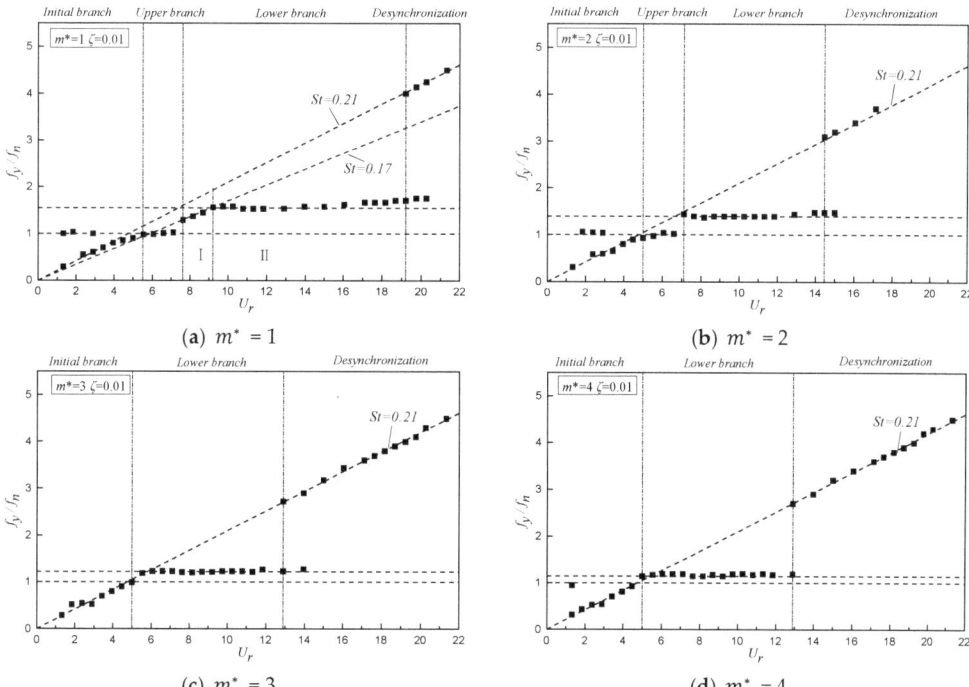

Figure 9. The frequency ratio (f_y/f_n) of the cylindrical oscillator under different mass ratios changes with Ur.

Figure 9 shows that, in the initial branch, cylindrical oscillators with mass ratios of one and two have an upper branch and the corresponding f_y/f_n is locked near one. However, cylindrical oscillators with mass ratios of three and four have only the initial branch and the lower branch, but no upper branch, and the corresponding f_y/f_n is not locked near one [26]. For cylindrical oscillators with a mass ratio of one, a transition region appears between the upper branch and the lower branch, in which $S_t \approx 0.17$, as shown in Figure 9a. However, this transition region does not appear in the frequency ratio curves of cylindrical oscillators with other mass ratios. In the lower branch, different mass ratios correspond to different f_y/f_n, and it tends toward one with the increase in mass ratio.

From Figure 9, it can be seen that in the initial branch, the cylindrical oscillator with mass ratios of one and two appears as the upper branch, and the corresponding f_y/f_n is locked near one; meanwhile, the cylindrical oscillator with mass ratios of three and four has only the initial branch and the lower branch without the upper branch, and the corresponding f_y/f_n is not locked near one. The cylindrical oscillator with a mass ratio of one has a transition region between the upper branch and the lower branch, and the Strouhal number in this region is 0.17, as Figure 9a shows. However, this transition region does not appear in the frequency ratio curves of other mass-ratio cylindrical oscillators. In the lower branch, the f_y/f_n corresponding to different mass ratios is different and tends toward one as the mass ratio increases.

It can be seen that in the lower branch, the smaller the mass ratio of the cylindrical oscillator, the larger the corresponding f_y/f_n value, and the cylindrical oscillators with mass ratios one and two will have higher transverse motion frequencies than the cylindrical oscillators with mass ratios of three and four.

5.5. Downstream Equilibrium Position

Figure 10 shows the variation in the equilibrium position of cylindrical oscillators with different mass ratios along the streamwise direction as a function of the reduced velocity. It can be seen from the figure that in the transition region between the upper branch and the lower branch of the cylindrical oscillator with a mass ratio of one, the equilibrium position in the streamwise direction suddenly decreases; it does not increase with the increase in the reduction speed but decreases initially and then increases. At the same reduction speed, the difference between the equilibrium positions with mass ratios of one and two is significantly larger than that with mass ratios of three and four, and this phenomenon is more obvious at higher reduction speeds. In the lower branch, the adventitial equilibrium position changes with the reduction speed [11].

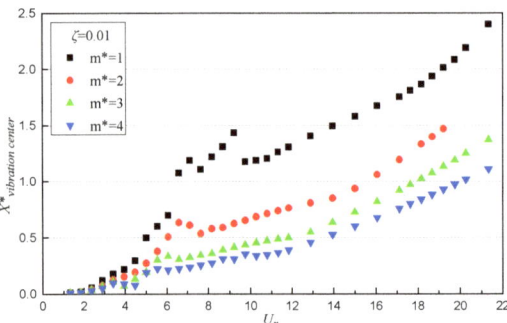

Figure 10. Variation in Ur in streamwise direction equilibrium position with different mass ratios.

Figure 10 gives the variation law of the downstream equilibrium position of cylindrical oscillators with different mass ratios with reduced velocity. As can be seen from the figure, the equilibrium position of the cylindrical oscillator with a mass ratio of one in the transition region between the upper branch and the lower branch shows a sudden decrease in the downstream equilibrium position, which does not increase with the increase in the reduced velocity, but first decreases and then increases, and the difference between the equilibrium positions of the cylindrical oscillators with mass ratios of one and two is significantly larger than that between the equilibrium positions of the cylindrical oscillators with mass ratios of three and four at the same reduced velocity. This phenomenon is more obvious at the higher reduced velocity [27].

The downstream equilibrium position increases with the reduced velocity, and at the same reduced velocity, the lower the mass ratio, the greater the corresponding downstream equilibrium position.

6. Summary and Conclusions

In this paper, the CFD numerical simulation method was used, the Fluent solver was applied, the SST k-ω turbulence model was selected, the SIMPLEC algorithm for pressure–velocity coupling in the momentum equation was applied and combined with the dynamic grid technique to study the effect of different mass ratios on vortex-induced vibration characteristics of the cylindrical oscillator of the hydrodynamics of a wind–wave combined power-generation system. This study provides theoretical support and a factual basis for our future research on the wind power generation part of the wind–wave combined power-generation system, for example, on the effect of the cylinder's roughness, the wind speed, or other parameters. Subsequent research will focus on the difference between 3D numerical models and 2D models of wind–wave combined power generation devices and on the cross-scale transfer of energy methods.

Additionally, the following conclusions are drawn after the study of the vortex-induced vibration power generation part of the wind–wave combined power generation system:

(1) Using the numerical simulation method compared with the classic experiment by Jauvtis and Williamson in 2004, the vortex-induced vibration of the cylindrical oscillator was well simulated, but the maximum amplitude was underestimated. The movement trajectories of the initial branch, the upper branch, and the lower branch are approximately the same as the experimental results. The reliability of the experiment is therefore verified.

(2) An analysis of the vortex vibration characteristics of cylindrical oscillators with mass ratios of one, two, three, and four reveals that with the increase in the reduced velocity, both the downstream amplitude and the transverse amplitude tend to decrease, and at the same reduced velocity, the overall trend is that the smaller the mass ratio, the larger the transverse amplitude, the more obvious the effect on the vortex vibration generator, and the higher the energy acquisition efficiency, but at the upper branch, the downstream vortex motion of the cylindrical oscillator is shown. However, in the upper branch, the crossflow vortex motion of the cylindrical oscillator "consumes" the energy of the crossflow vortex motion, and the mass ratio of one is significantly larger than the mass ratio of two, making the crossflow amplitude of the cylindrical oscillator with mass ratio one smaller than the mass ratio of two.

(3) In the frequency ratio f_y/f_n curve, the cylindrical oscillator with a mass ratio of one has a transition region between the upper branch and the lower branch but not in other mass ratio cylindrical oscillators. For different mass ratio cylindrical oscillators, f_y/f_n tends toward one continuously as the mass ratio increases, and in the lower branch, the downstream equilibrium position varies with the reduced velocity, and the lower the mass ratio, the larger the corresponding downstream equilibrium position at the same reduced velocity.

Author Contributions: Writing—original draft, H.S.; Writing—review & editing, J.W.; Supervision, H.L., G.H., Z.Z., B.G. and B.J. All authors have read and agreed to the published version of the manuscript.

Funding: This research was funded by the National Natural Science Foundation of China Youth Fund (grant number 51909148).

Conflicts of Interest: The authors declare no conflict of interest.

References

1. Alam, M. Effects of mass and damping on flow-induced vibration of a cylinder interacting with the wake of another cylinder at high reduced velocities. *Energies* **2021**, *14*, 5148. [CrossRef]
2. An, X.; Song, B.; Tian, W.; Ma, C. Design and CFD simulations of a vortex-induced piezoelectric energy converter (VIPEC) for underwater environment. *Energies* **2018**, *11*, 330. [CrossRef]
3. Lv, Y.; Li, J.; Pan, S.; Wang, W. Vortex Induced Vibration Generator and Its Key Technology. *Equip. Manuf. Technol.* **2020**, *1*, 161–165.
4. Zhou, B.; Hu, J.; Xie, B.; Ding, B.; Xia, Y.; Zheng, X.; Lin, Z.; Li, Y. Research Progress in Hydrodynamics of Wind-Wave Combined Power Generation System. *Chin. J. Theor. Appl. Mech.* **2019**, *51*, 1641–1649.
5. Sun, H.; Huang, W.; Li, L.; Chang, S.; Hu, F. Experimental Study of Vertex Induced Motion of Spar-Type Floating Support Structure Forwind Turbine. *Acta Energ. Sol. Sin.* **2017**, *38*, 3412–3418.
6. Li, L.; Tan, D.; Yin, Z.; Wang, T.; Fan, X.; Wang, R. Investigation on the multiphase vortex and its fluid-solid vibration characters for sustainability production. *Renew. Energy* **2021**, *175*, 887–909. [CrossRef]
7. Yu, J.; Li, Z.; Yu, Y.; Hao, S.; Fu, Y.; Cui, Y.; Xu, L.; Wu, H. Design and Performance Assessment of Multi-Use Offshore Tension Leg Platform Equipped with an Embedded Wave Energy Converter System. *Energies* **2020**, *13*, 3991. [CrossRef]
8. Bernitsas, M.M.; Raghavan, K.; Ben-Simon, Y.; Garcia, E.M.H. VIVACE (vortex induced vibration aquatic clean energy): A new concept in generation of clean and renewable energy from fluid flow. *Int. Conf. Offshore Mech. Arct. Eng.* **2006**, *47470*, 619–637.
9. Jauvtis, N.; Williamson, C.H.K. The effect of two degrees of freedom on vortex-induced vibration at low mass and damping. *J. Fluid Mech.* **2004**, *509*, 23–62. [CrossRef]
10. Feng, C.C. The Measurement of Vortex Induced Effects in Flow Past Stationary and Oscillating Circular and D-Section Cylinders. Doctoral Dissertation, University of British Columbia, Vancouver, BC, Canada, 1968.
11. Huang, Z.; Pan, Z.; Cui, W. Numerical simulation of VIV of a circular cylinder with two degrees of freedom and low mass-ratio. *J. Ship Mech.* **2007**, *11*, 1–9.
12. Du, X.; Tang, C.; Zhao, Y.; Wu, G.; Yang, X. Effects of mass ratio on the vortex-induced vibration of two types of tandem circular cylinders. *J. Vib. Shock* **2022**, *41*, 160–168.

13. Bai, X. Analysis of the parameters influencing the ocean current energy harvesting efficiency of circular cylinder VIV. *Renew. Energy Resour.* **2017**, *35*, 784–790.
14. Tan, J.; Wang, B.; Yuan, P.; Wang, S.; Chen, C.; Zheng, Z. Study on Turbulence Intensity Influence on Cylindrical Oscillator Response of VIV Tidal energy Conversion Device. *Acta Energ. Sol. Sin.* **2020**, *41*, 20–26.
15. Khalak, A.; Williamson, C.H.K. Dynamics of a Hydroelastic Cylinder with Very Low Mass and Damping. *J. Fluids Struct.* **1996**, *10*, 455–472. [CrossRef]
16. Yang, X.; Zhao, Y.; Du, X.; Wu, G. Effects of mass ratio on wake-induced vibration of two tandem circular cylinders and its mechanism. *J. Vib. Eng.* **2020**, *33*, 24–34.
17. Han, C. Analysis and Prediction Reasearch on Energy Harvesting Effciency of Flow-Induced Vibration with Two Tandem Vibrators Containing Roughness. Master's Thesis, Jiangsu University of Science and Tecnology, Zhenjiang, China, 2020.
18. Shan, X.; Tian, H.; Cao, H.; Xie, T. Enhancing performance of a piezoelectric energy harvester system for concurrent flutter and vortex-induced vibration. *Energies* **2020**, *13*, 3101. [CrossRef]
19. Pan, Z.Y.; Cui, W.C.; Miao, Q.M. Numerical simulation of vortex-induced vibration of a circular cylinder at low mass-damping using RANS code. *J. Fluids Struct.* **2007**, *23*, 23–37. [CrossRef]
20. Gao, Y.; Zong, Z.; Zou, L.; Takagi, S.; Jiang, Z. Numerical simulation of vortex-induced vibration of a circular cylinder with different surface roughnesses. *Mar. Struct.* **2018**, *57*, 165–179. [CrossRef]
21. Hao, W.; Dapeng, S. Study on suppresstion measures for Vortex-Induced Vibration of the Deep Water Riser. *China Offshore Platf.* **2009**, *24*, 1–8.
22. Meneghini, J.R.; Saltara, F.; Fregonesi, R.A.; Yamamoto, C.T. Vortex-induced vibration on flexible cylinders. In *Numerical Models in Fluid-Structure Interaction*; WIT Transactions on State-of-the-art in Science and Engineering; WIT Press: Billerica MA, USA, 2005; Volume 18.
23. Zhang, W.; Li, S.; Liu, Y.; Li, D.; He, Q. Optimal Control for Hydraulic Cylinder Tracking Displacement of Wave Energy Experimental Platform. *Energies* **2020**, *13*, 2876. [CrossRef]
24. Gonçalves, R.T.; Rosetti, G.F.; Franzini, G.R.; Meneghini, J.R.; Fujarra, A.L.C. Two-degree-of-freedom vortex-induced vibration of circular cylinders with very low aspect ratio and small mass ratio. *J. Fluids Struct.* **2013**, *39*, 237–257. [CrossRef]
25. Zdravkovich, M.M. Flow induced oscillations of two interfering circular cylinders. *J. Sound Vib.* **1985**, *101*, 511–521. [CrossRef]
26. Gu, J.; Yang, C.; Zhu, X.; Wu, J. The influence of mass ratio on the vortex induced motion characteristics of circular cylinder. *J. Vib. Shock* **2016**, *35*, 134–140.
27. Bai, X.; Han, C.; Cheng, Y. Parametric Analysis of an Energy-Harvesting Device for a Riser Based on Vortex-Induced Vibrations. *Energies* **2020**, *13*, 414. [CrossRef]

Article

Deep Learning for Modeling an Offshore Hybrid Wind–Wave Energy System

Mahsa Dehghan Manshadi [1], Milad Mousavi [1,2], M. Soltani [1,3,4,*], Amir Mosavi [5,6,*] and Levente Kovacs [7,8]

1. Department of Mechanical Engineering, K. N. Toosi University of Technology, Tehran 1999143344, Iran
2. Faculty of Informatics, Selye University, 94501 Komarom, Slovakia
3. Department of Electrical and Computer Engineering, University of Waterloo, Waterloo, ON N2L 3G1, Canada
4. Waterloo Institute for Sustainable Energy (WISE), University of Waterloo, Waterloo, ON N2L 3G1, Canada
5. German Research Center for Artificial Intelligence, 26129 Oldenburg, Germany
6. Institute of Information Engineering, Automation and Mathematics, Slovak University of Technology in Bratislava, 81243 Bratislava, Slovakia
7. Biomatics and Applied Artificial Intelligence Institution, John von Neumann Faculty of Informatics, Obuda University, 1034 Budapest, Hungary
8. Physiological Controls Research Center, Obuda University, 1034 Budapest, Hungary

* Correspondence: msoltani@uwaterloo.ca (M.S.); amir.mosavi@uni-obuda.hu (A.M.)

Citation: Dehghan Manshadi, M.; Mousavi, M.; Soltani, M.; Mosavi, A.; Kovacs, L. Deep Learning for Modeling an Offshore Hybrid Wind–Wave Energy System. *Energies* **2022**, *15*, 9484. https://doi.org/10.3390/en15249484

Academic Editor: Rosa Anna Mastromauro

Received: 30 September 2022
Accepted: 23 November 2022
Published: 14 December 2022

Publisher's Note: MDPI stays neutral with regard to jurisdictional claims in published maps and institutional affiliations.

Copyright: © 2022 by the authors. Licensee MDPI, Basel, Switzerland. This article is an open access article distributed under the terms and conditions of the Creative Commons Attribution (CC BY) license (https://creativecommons.org/licenses/by/4.0/).

Abstract: The combination of an offshore wind turbine and a wave energy converter on an integrated platform is an economical solution for the electrical power demand in coastal countries. Due to the expensive installation cost, a prediction should be used to investigate whether the location is suitable for these sites. For this purpose, this research presents the feasibility of installing a combined hybrid site in the desired coastal location by predicting the net produced power due to the environmental parameters. For combining these two systems, an optimized array includes ten turbines and ten wave energy converters. The mathematical equations of the net force on the two introduced systems and the produced power of the wind turbines are proposed. The turbines' maximum forces are 4 kN, and for the wave energy converters are 6 kN, respectively. Furthermore, the comparison is conducted in order to find the optimum system. The comparison shows that the most effective system of desired environmental condition is introduced. A number of machine learning and deep learning methods are used to predict key parameters after collecting the dataset. Moreover, a comparative analysis is conducted to find a suitable model. The models' performance has been well studied through generating the confusion matrix and the receiver operating characteristic (ROC) curve of the hybrid site. The deep learning model outperformed other models, with an approximate accuracy of 0.96.

Keywords: renewable energy; artificial intelligence; machine learning; comparative analysis; wind turbine; energy; deep learning; big data; wave energy; wave power; offshore

1. Introduction

In recent years, a significant part of the energy conversion mechanism in renewable energy systems (RES) has utilized the ocean's waves energy. Energy harvesting from the oceans was an efficient and clean way of producing electricity [1]. This relatively new energy resource can significantly reduce the pressure on fossil fuel power plants and positively contribute to reducing carbon dioxide emissions and further pollutants [2]. Recently, there has been a great deal of progress in advancing the energy conversion mechanisms for renewable energy systems (RESs) [3,4]. Hybrid RESs, e.g., wave–wind combined systems, have also emerged to improve efficiency and performance [5,6]. Rony and Karmakar investigated the integrated system's responses to understand the effects of the wave energy converter (WEC) in the various operating conditions on the wind turbine under regular and irregular waves. This study presents a suitable array of these systems in a hybrid RES site [7]. Another research is implemented on the other aspect

of these kinds of sites. Si et al. studied the dynamic response and output power of the float wind turbine and wave energy converter. As a result, the optimum array of these systems and their best control design are introduced [8]. In order to use a specific area to generate optimum electricity, researchers have found a way to combine RESs that can simultaneously generate optimized electricity from multiple energy sources. They found that the ocean environment could cause a significant amount of electricity from a combination of wind turbines and wave energy converters due to sufficient wind currents and naturally occurring waves in these environments [9,10]. It has been well studied that optimal use of offshore wave–wind energy reduces the emitted amount of CO_2 into the atmosphere and contributes to the region's economic growth and development [11,12]. Today, the simultaneous use of wind and ocean waves can be considered a well-established practice and essential for developing the new generation of offshore energy farms [13–15]. The wind–wave hybrid system has also been regarded as a sustainable RES [16] for the optimal use of clean and free energy sources [17] and low-cost production [18,19] with lower environmental impacts [20,21]. Related to the importance of increasing the demand rates of these sites, this study contributes to the optimal design and advancement of a wave–wind system using a data-driven method [7,22]. Consequently, in this proposed research, hybrid bladeless wind turbines and wave energy converters, the simultaneous use of wind and wave energy, is considered as the hybrid site of the bladeless wind turbine and the wave energy converter (HBWTWEC) [23]. Despite the high installation costs of these sites, the construction procedure can have many benefits for private and nonprivate investors, such as reducing construction costs through electrical energy transfer, storage systems, common infrastructure [24], and increasing the amount of produced energy in the specified area [25,26]. Hence, it is clear that power plants must have specific guidelines for operation due to the inherent characteristics of the systems and their performance in the harsh conditions of oceans [27]. For the mentioned reasons, the main factor for the technological progress of these sites can be referred to as the global development of numerical simulations and appropriate cost-effective models [28]. These models can provide appropriate facilities for evaluating these sites according to the installation area. The recent studies on hybrid RES can be observed in Table 1.

Table 1. Recent studies on hybrid RES sites by combining wave energy converter and wind turbine.

	Authors	Concept	Year	Method	Description
1	Mohammad Hossein Jahangir, et al. [29]	Zero-emission PV/Wind turbine/Wave energy converter	2020	A techno-economic and environmental analysis for a hybrid renewable energy system	Feasibility study of wave energy hybridization with solar, wind, and storage systems
2	Yu Zhou, et al. [30]	Wave energy converter integrated monopile	2020	Hydrodynamic investigation of hybrid renewable energy systems	The hydrodynamic efficiency of the OWC device decreases with the wave nonlinearity
3	Yulin Si, et al. [31]	Semi-submersible floating wind turbine and point-absorber wave energy converter	2021	Power take-off controls are implemented for hybrid renewable energy system	A novel hybrid floating wind and wave power generation platform is proposed
4	A.H. SamithaWeerakoon, et al. [32]	Vertical augmentation crossflow turbine	2021	ANSYS-CFX optimized and evaluated both experimentally and computationally.	A novel vertical augmentation channel, with nozzles on both sides of the turbine, was designed, and an optimized configuration was obtained and evaluated as a wave energy converter.

Among the industrial tools for data-driven modeling, different machine learning (ML) and deep learning (DL) algorithms, such as various types of artificial neural networks (ANN), in addition to the Internet of Things (IOT) technologies, are found to be efficient for modeling and forecasting offering an alternative way to solve complex problems [33–35]. In particular, DL algorithms can be trained with the datasets obtained from numerical simulations by various computational fluid dynamics (CFD) methods, as well as experimental data from laboratories [36]. Deep learning algorithms have been used to predict essential parameters and achieve gradual and more accurate performance. Furthermore, they are considered to be an excellent tool for predicting energy systems whose inputs are variable in nature. The atmospheric parameters are unstable, which can affect the system's output [37–40]. Among the applications of this method in wind energy systems, we can mention research on rapid and accurate forecasting of wind speed during a day, month, and even over several years as a climate change model [37,39]. These prediction methods were also mentioned in analyzing the output power of different wave energy converters. Many studies have been conducted in this field, which is evidence of how artificial intelligence can present wave height as a function of wind speed and predict the efficiency of these converters [41,42]. Similarly, ML-based prediction methods have been used in further hybrid renewable energy systems, and it is expected that the complex characteristics of the hybrid sites can be easier predicted and improved. Therefore, this study proposes a novel concept for a hybrid wind-wave energy converter, where the Searaser is considered for the WEC sub-system integrated with a novel vortex bladeless turbine. Mousavi et al. [41] investigated the numerical solution model using experimental data and predict the amount of production power using the LSTM method. Moreover, Dehghan et al. [42] simulated the prototype turbine built experimentally in laboratory conditions and estimated its production power. The research gap is the urge for more evaluations in the real case scenario of a hybrid wind-wave energy in a marine power plant. This study comprehensively compares the efficiency of two energy systems with a specific input, which is the experimental data used to tackle the first part of the development of the hybrid system. Furthermore, two systems are integrated into an offshore power plant for a specific location. The data-driven methods are used to predict the power generation of the two cases, which have a vital role in simultaneously harvesting energy from wind and wave and alleviating investment risk.

Due to the remote coastal location of the experimental test and climate change modeling, it is essential to use wind turbines and wave energy converters due to the potential renewable energy resources of these regions. Alternative renewable energy systems, e.g., solar cannot perform efficiently in the region. This study brings novelty by investigating the combination of a special kind of wind turbine and wave energy converter, namely vortex bladeless wind turbine (VBT) and Searaser. Numerical simulation was selected as an input in the case of a prerequisite for providing data, beginning with selecting a location where the experimental data were collected. For this purpose, the results and experimental test values are used as the ML method's input. The main aim is to utilize different ML methods to accurately predict desired parameters by the most common RES. They are recurrent neural networks (RNNs), long short-term memory (LSTM), random forest, and support vector machine (SVM), which were applied to the same input. Hence, the training procedure of these algorithms requires appropriate data. Moreover, their forecasting performance is compared between these different algorithms. In addition, the output power for each system in a hybrid power plant is calculated and compared, and finally, the output power of the hybrid site is calculated, respectively.

2. Materials and Methods

In this section, the introduced hybrid site includes the array of two popular RESs: a wind turbine and wave energy converter. Due to the importance of solving real-world problems, the dataset is collected from the experimental test. Then, by utilizing the input dataset from environmental conditions, different ANN algorithms are developed to predict

the net produced power of the hybrid site. Figure 1 presents the different steps of this study.

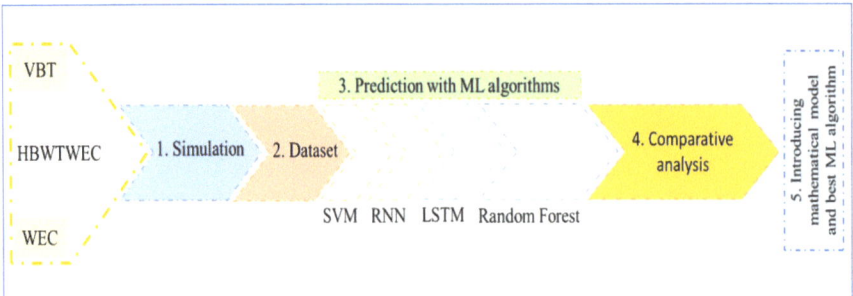

Figure 1. Overview of the proposed study.

2.1. HBWTWEC Description

The hybrid site includes a combination of offshore wind and wave energy systems. The proposed wind turbine of this research is a vortex bladeless wind turbine (VBT), and the wave energy converter is a Searaser. Using a large number of these systems, the offshore site can be built to generate a significant amount of electrical energy. In this research, it consists of ten VBTs and ten Searasers. In the following, the forecasting models are studied and compared, along with finding the best system among the two proposed RESs in terms of power generation.

2.2. Experimental Data

In order to make the results more realistic, the experimental data from a climate change model [43] were used as the ML algorithm's input to simulate and solve the governing equations numerically. For this purpose, the proposed parameters are defined as wind speed and wave height as input conditions. The data variate abruptly because of their fluctuated nature, which changes with the local weather [43].

2.3. Numerical Simulations

One of the most important points, in order to have accurate artificial intelligence (AI) prediction modeling, is to have a rich dataset. Two critical parameters are collected from an experimental test. However, for hybrid site modeling, more input features are required. In this case, it is necessary to have numerical modeling first.

Governing Equations

To simulate HBWTWEC, numerical solution software (FLOW-3D) was used to analyze solid and fluid interactions between the structure of a VBT and airflow, as well as Searaser and ocean waves. This software utilizes the volume fraction technique as a computational cell to calculate the ratio of open volume to total in a computational volume [44]. The study is classified into two different systems, where the involved fluids are different, respectively.

A. Searaser and ocean waves interaction equations.

The exerted force from the buoy which makes a torque is explained by Equation (1) [41].

$$\vec{F} = m \frac{d\vec{v_G}}{dt} \qquad (1)$$

where m is the mass of the buoy and $\frac{d\vec{v_G}}{dt}$ represents the acceleration of the buoy, which is derived from the velocity of the buoy relative to time (t). V indicates the speed at which the buoy moves along the proportional axis to the ocean's surface (z). The torque helps

to produce mechanical power. This motivates a special generator to convert this power to electrical power [41].

$$P = \frac{1}{64\pi}\rho_s g^2 H_s^2 T \qquad (2)$$

where ρ_s is the density of seawater and g is the acceleration of the Earth's gravity. H_s is the wave height passed from the Searaser, and T is the torque of the buoy's rotational movement.

B. Interaction equations of bladeless wind turbine and airflow.

The equation of drag force exerted on the body of the VBT and the generated power were solved to produce a related dataset. Other governing equations are given in detail in [43]. Equations (3) and (4) show these two important parameters in this research.

$$F_{fluid}(x.y.t) = \frac{1}{2}\rho u^2 (Dl) C_d(x.y) \sin(\omega t + \varphi)\hat{i} + \frac{1}{2}\rho v^2 (Dl) C_d(x.y) \sin(\omega t + \varphi)\hat{j} \qquad (3)$$

where ρ is the air density, u is the wind speed, D is the diameter of the turbine oscillator, and l is the body's height. The drag coefficient (C_d) depends on the x and y axes, and $\sin(\omega t + \varphi)$ is the sine oscillating with angular velocity ω and phase difference φ [43]. This equation indicates that this converted force is a special drag force which is directly related to the VBT geometry and the flow properties.

$$P = \eta \frac{1}{2}\rho U^3 (2y + D)l \qquad (4)$$

Since η represents the energy conversion factor of VBT and y is the amplitude of VBT oscillation, that should be considered as a variable in further calculations. The produced power of VBT has a drag force nature, too. It highly depends on the flow velocity (wind speed) and properties and also the VBT geometry, respectively. Unlike the interaction of these two systems on each other, which has been investigated, this study has neglected this interaction. Conventional offshore wind turbines not only move the surrounding air with a rotational movement of blades but also affect the sea waves. As a result, they will affect the performance of the wave energy converter. On the other hand, VBT has not affected Searaser performance due to vibrational movement, with only very small amplitudes that can be ignored at that scale.

2.4. Dataset Preparation

Due to the database nature of ML methods, it is necessary to provide a database for the algorithms to predict the required parameters. Therefore, to give a large amount of data, we use simulation and numerical solutions of governing equations and use their output as a dataset needed for training. The type of experimental training data is segmented by the splitting method, and its ratio is 90 to 10. The 5% of the dataset is randomly selected for the evaluation.

2.5. Machine Learning Algorithms

In this study, in addition to examining HBWTWEC and comparing the bladeless wind turbine with Searaser in the case of production capacity, a comparative analysis using different ML methods was conducted to find the best method. The best one with the highest efficiency predicts the total produced power in an RES site in a specific area. The utilized methods can be LSTM, RNN, random forest, and SVM, which are examined in the following governing equations of each algorithm.

2.5.1. RNN and LSTM

The Recurrent neural network (RNN) is often used to model the data for identifying each sample as dependent on previous samples, and the convolution layers extend the neighborhood to the desired pixels. Despite its advantages, it has problems with gradients disappearing and exploding during calculations, and the process of teaching this algorithm

is a bit more complicated than other algorithms. Moreover, during processing, if one of the "tanh" or "Relu" activation functions is used, it can no longer perform sequential processing [45]. The optimized LSTM algorithms are a type of recurring neural network that facilitates the storage of data in memory. On the other hand, this algorithm solves the problem of gradient disappearance in the RNN algorithm. It should be noted that this algorithm has excellent performance for different stages of classifying, processing, and predicting time series.

1. The first gate, as expected, is the input gate, which decides how many inputs should be used for the algorithm's memory change operation. The activation function in these two algorithms is the Sigmoid function, which decides which values to pass according to the two choices of 0.1. In addition, the "tanh" function is used to weight the input values, and this function's output range varies from -1 to 1.
2. The second gateway in this algorithm is the forget-me-not gateway, which argues which data should be removed from this process within each block and how this gateway works like an input valve with a sigmoid function. This gateway compares the data of the previous state and the data recently entered into the block and shows the number 1 or 0 for each datum in that cell. The zero indicates that the data should be forgotten there. Additionally, 1 means that the data should be stored and used to continue the process.
3. The last gate, as expected, is the exit gate. Like the other two gates, the Sigmoid function decides which values to pass through 0.1. Additionally, the tanh function, like the input gate, weighs these values from -1 to 1.

The hyper parameters of RNN are 10 hidden layers, with Adam activation and 1 dense layer. The epoch size is 100 and the batch size is 20. Moreover, for the LSTM algorithm, the hyper parameters are the same as RNN in order to have a better comparison analysis between these two algorithms.

2.5.2. Random Forest

One ensemble method of ML can be introduced as random forests (RF) or random decision forests, which are often trained for regression and classification, which works by building a large number of decision trees during training. In this study, the regression prediction of this algorithm is considered more than its classification, and this is such that the average prediction of each tree in the whole forest, which is a set of decision trees, is returned. After several rounds of training, random decision forests accustom the existing trees according to the data and their change tags to perform better [46].

2.5.3. SVM

This algorithm is used to categorize and regress data. In this study, because the data are scattered and not necessarily linear, it is better to use this algorithm to predict values. In the algorithm, we plot each datum as a point in multidimensional space, the number of dimensions of which is equal to the number of properties in the problem under study. Then, the algorithm performs the classification by finding a surface that connects the two features well, and the abundance of data on that page is higher [47]. One of the most important equations in the study [38] is the equation of applied force to the buoy and the produced power of the Searaser. Equations (1) and (2) are used in order to achieve one of the goals [42].

3. Results

Since this study is conducted to promote the view of the ML application based on estimating the production capacity of a hypothetical power plant, the material presented as a simulation and numerical solution has been used in the previous two studies. Furthermore, valuation of results has been performed in previous articles. Figure 2 shows the evaluation and validation of the results by comparing the amount of buoyancy in the vertical axis. The value of the difference between the two graphs is due to the difference in

the simulation input. In a former study by Babajani et al. [46], this value was a maximum of 0.75 m, but in this study, the maximum height of the input wave is 1 m.

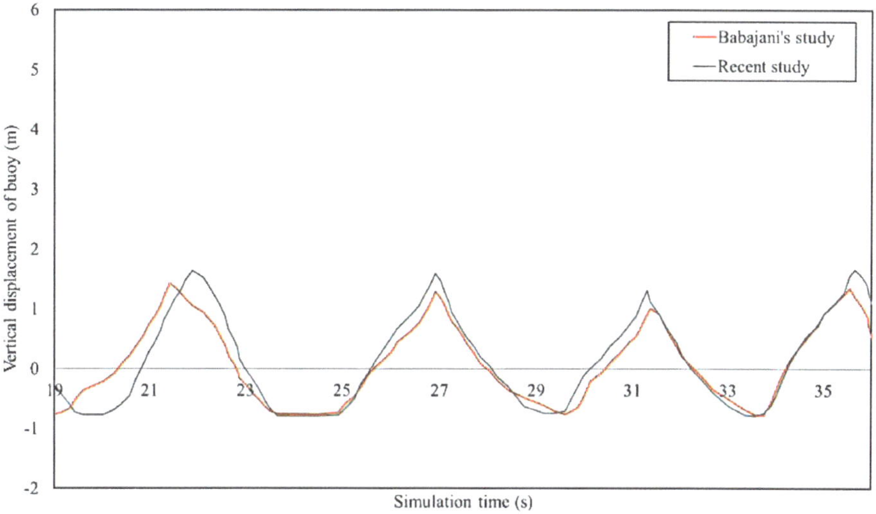

Figure 2. Validation of recent work with the former study.

As it is observed in Figure 2, results are so close to another study, which is evidence of verification. As a matter of fact, the vertical displacements of the buoy fluctuate 3 m. It is moved up to 2 m up and 1 m down from its origin. These periodic movements happen during simulation time. The periodic nature of buoy movement is because of the periodic behavior of inlet waves.

3.1. Comparative Analysis between Different Machine Learning Algorithms

Since one of the most important issues in the analysis of RESs for the construction of hybrid sites is to predict the amount of produced power of these systems, in this study, we have tried to find the best one among the four mentioned algorithms. Therefore, this study was conducted to find the best ML algorithm in the most practical parameters used in system analysis as key parameters in maintaining systems and selecting them to build an RES site, the force exerted on each of the systems, and their produced power. Figures 3 and 4 show the ML algorithms predicted results for these two parameters.

Figure 3 presents the converted force of the VBT (Figure 4a) and WEC (Figure 4b). The nature of the exerted force in VBT is drag force. So, it varies with power law. However, WEC profiles fluctuated as a sine function because they are completely depended on ocean waves as an inlet of this system. The VBT maximum forces are 4 kN, and for the WEC are 6 kN, respectively. Moreover, the values predicted by various machine learning methods (LSTM, SVM, RNN, RF) are compared with the amount of exerted force on the bladeless wind turbine (a) and Searaser (b). It can obviously be seen that the best algorithms among them are LSTM and RNN, which have the closest prediction values to the simulation ones. Moreover, it shows that the mentioned algorithms are more reliable than others. However, it can be realized that the best one is the LSTM algorithm. The advantage of this method over RNN is expected to be the solution to the sudden disappearance problem of gradients during code execution.

Figure 4 also presents a comparative analysis, considering that the LSTM algorithm is the best way to predict the output power of these two systems. In these predictions, the only algorithm that did not work accurately is the random forest algorithm.

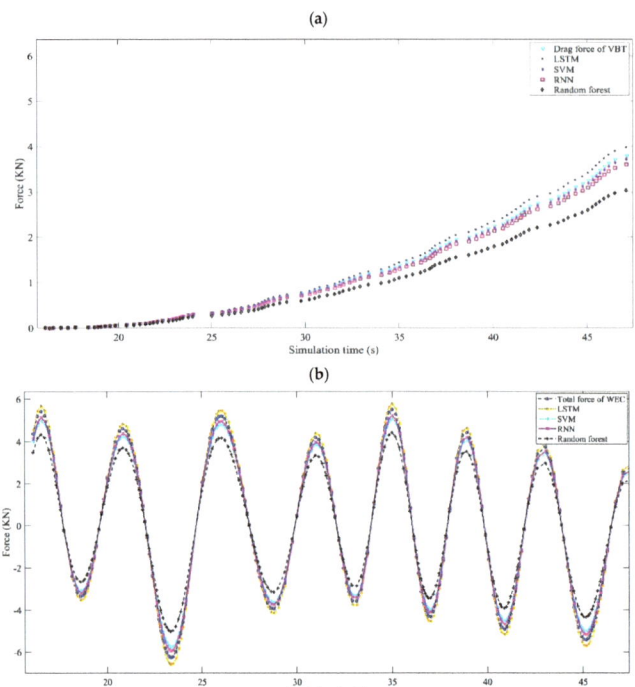

Figure 3. Forces over 50 s of simulation for (**a**) VBT and (**b**) WEC.

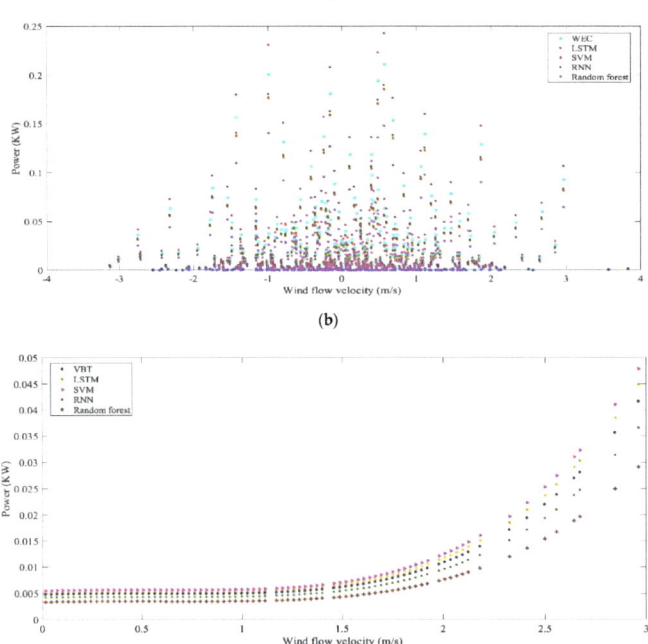

Figure 4. Power produced over 50 s of simulation for (**a**) VBT and (**b**) WEC.

3.2. Statistical Analysis for Evaluating Model Performance

One of the most important things in ML implementation is to evaluate these algorithms with the ROC curve. Furthermore, it is another tool to examine the accuracy of their performance. Three commonly utilized ML algorithms for estimating the power and also converted force were assessed, and the model performances are reported in Table 2. Statistical analysis was then assessed using mean absolute error (MAE), root mean square error (RMSE), ACC, FPR, TPR, PPV, and TNR. Furthermore, ROC curves and the confusion matrix of different values related to each parameter are shown in Figure 5.

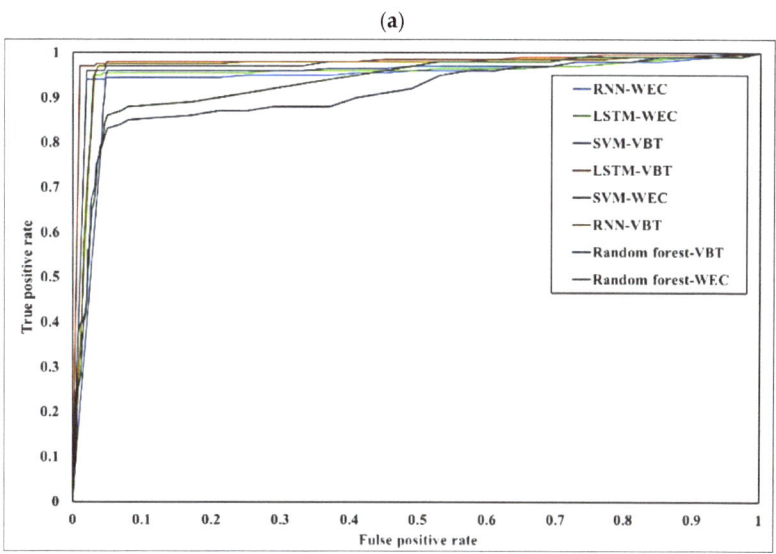

Figure 5. ROC curve (**a**) and confusion matrix (**b**) of different utilized algorithms in predicting VBT and WEC values.

Table 2. Evaluation of machine learning parameters.

Parameter	Method	TNR	PPV	TPR	FPR	ACC	RMSE	MAE
WEC Force	RNN	0.950	0.951	0.924	0.050	0.937	32.057	0.066
	LSTM	0.978	0.980	0.942	0.022	0.959	16.422	0.064
	SVM	0.980	0.981	0.903	0.020	0.938	25.874	0.069
	RF	0.913	0.918	0.874	0.087	0.892	31.984	0.067
BWT Force	RNN	0.960	0.961	0.933	0.040	0.946	24.885	0.032
	LSTM	0.980	0.980	0.916	0.020	0.946	37.526	0.068
	SVM	0.966	0.969	0.896	0.034	0.928	15.425	0.071
	RF	0.951	0.950	0.941	0.049	0.946	27.344	0.036
WEC Power	RNN	0.959	0.960	0.941	0.041	0.950	15.535	0.032
	LSTM	0.970	0.970	0.906	0.030	0.936	16.422	0.064
	SVM	0.978	0.979	0.901	0.022	0.937	15.874	0.069
	RF	0.987	0.988	0.898	0.013	0.939	31.984	0.067
BWT Power	RNN	0.942	0.942	0.915	0.058	0.928	24.885	0.032
	LSTM	0.961	0.961	0.907	0.039	0.933	37.523	0.068
	SVM	0.968	0.969	0.931	0.032	0.949	15.420	0.071
	RF	0.982	0.983	0.897	0.018	0.937	27.340	0.036

Figure 5 shows that the most accurate algorithm, as it can be considered, is LSTM. When tested for four proposed ML algorithms, the significant accuracy is for the LSTM algorithm in four measured parameters. The true positive rates of this algorithm are in the highest level in power and exerted forces for both proposed renewable energy systems. Moreover, the false negative rates are the least in predicting these desired parameters. However, RNN is another accurate algorithm for predicting these parameters, but related to this study's aim, the most accurate one should be introduced.

3.3. Comparison between WEC and BWT

Another purpose of this study is to conduct analyses in order to select the best energy system from VBT and WEC in specific locations with proposed geographical conditions. Analyses include measuring the amount of applied force to each system and their output electrical power. Figure 6 compares the force from the waves with the WEC and the ocean airflow with the VBT.

Figure 6 shows the diagram of drag force from the airflow to the moving part of the VBT and the total force on the Searaser during the simulation time. Moreover, by fitting the curves of both graphs, the force on each system can be estimated separately. The drag force can be introduced as a quadratic function (Equation (5)), but the exerted force on the Searaser can be the summation of five sine functions (Equation (6)). Equations (5) and (6) represent the equation obtained from the curve fitting.

$$y = 0.0039x^2 - 0.13x + 1 \tag{5}$$

$$f(x) = a_1 \sin(b_1 x + c_1) + a_2 \sin(b_2 x + c_2) + a_3 \sin(b_3 x + c_3) + a_4 \sin(b_4 x + c_4) + a_5 \sin(b_5 x + c_5) \tag{6}$$

Different parameters of Equation (6) are shown in Table 3.

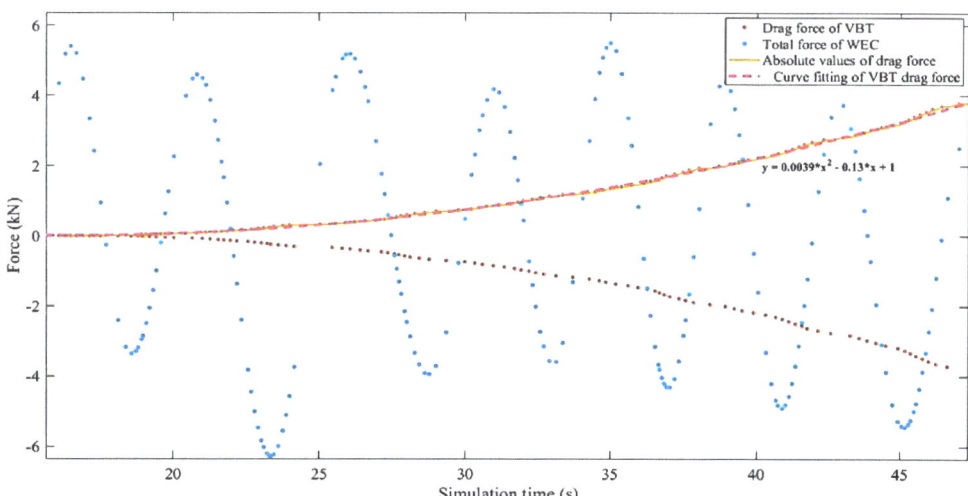

Figure 6. Drag and total force exerted on a VBT.

Table 3. Values of curve fitting parameters related to Equation (6).

Parameter	Value
a1	3921
a2	2063
a3	1585
a4	2061
a5	1779
b1	1.418
b2	1.255
b3	0.9509
b4	1.643
b5	0.8553
c1	2.8888
c2	−0.2617
c3	9.099
c4	0.2676
c5	2.915

Another investigation is to find the related mathematical equations of the proposed RES-produced power. Figure 7 presents the generated electrical power in each RES, respectively.

Moreover, Figure 7 shows the electrical power of each system with respect to energy conversion. For this diagram, curve fitting was performed to provide relationships to estimate their production power in terms of wind speed. The difference between the two diagrams in Figure 6 is due to the nature of the points drawn in the figure, no relation can be introduced to Searser's productivity. Equation (7) presents the produced power of VBT.

$$P = 0.0006v^4 - 0.00002v^3 - 0.0013v^2 + 0.0009v + 0.0048 \tag{7}$$

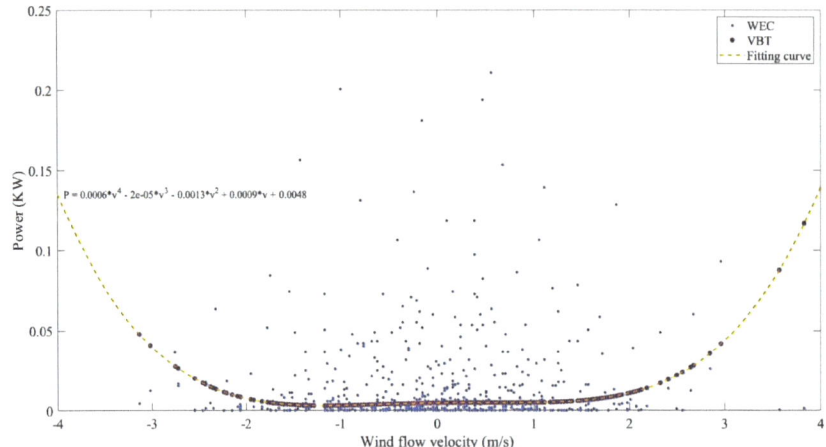

Figure 7. Generated power of WEC and VBT by the curve fitting.

3.4. Total Energy of HBWTWEC

Another important goal of this study is to obtain the total produced power in an HBWTWEC. Due to the nature of the generated electrical power of Searaser in terms of wind speed, a definite mathematical equation cannot be provided. Figure 8 shows the production capacity for each system, as well as the total produced power of HBWTWEC in a considered location.

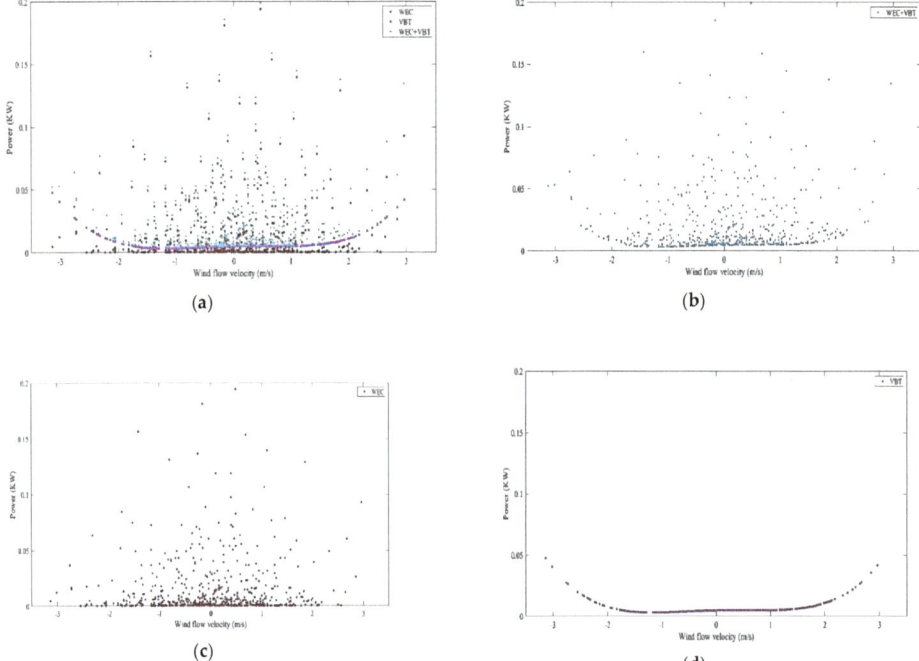

Figure 8. (**a**) Illustration of the energy generated by WEC and VBT, individually and combined: (**b**) Combined generated power of VBT and WEC (**c**) Generated power of WEC (**d**) Generated power of VBT.

This diagram helps to estimate the desired values completely, which can be used in order to construct a hybrid site according to the wind speed from upstream of the ocean (Figure 8a). Due to the scatter values of Searaser-produced power (Figure 8c), when it is summed by the VBT-produced power (Figure 8d), which has a specific equation (Figure 8b), it cannot be introduced as an obvious mathematical equation for modeling. For having a comparison between two RESs, it can be noticed that the highest value of power belongs to the Searaser. On the other hand, the minimum order of produced power is related to the VBT. For the future research, exploring the applications of the metaheuristics and further deep learning e.g., [47–49] methods are proposed.

4. Conclusions

In this study, HBWTWEC simulations consisting of ten VBTs and ten Searasers during the FETCH experiment are performed by Flow-3D software. Partial climate change models are used to simulate the region's local climate as an input of mathematical solution of governing equations. Then, not only the force on each system and the produced power are calculated separately but, also, they are compared to select the most suitable one according to the weather conditions of the selected region. The maximum and minimum values of the produced power belong to the Searaser and VBT, respectively. One of this study's most significant achievements is introducing a mathematical equation for two essential variables: the exerted force and produced power of introduced systems. These are measured by best fitting the output graphs from the numerical simulations. The drag force, which is exerted by the wind blowing across the VBT, is introduced as a quadratic function, and the total exerted force from the ocean's waves on Searaser is modeled as a summation of five sinusoidal functions. However, there has not been any known function for the produced power of HBWTWEC due to the scattered nature of Searaser's output power. Since the most important issue in the hybrid site development is the estimation of produced power by RESs installed in that area, high-precision algorithms must be evaluated in order to introduce the best of them. The most accurate algorithm can be introduced as LSTM. However, RNN is another accurate algorithm, but related to the significant goal of the study, the most accurate one is presented. These key predictions in the hybrid site's industry include the two parts of measuring the forces exerting on systems and their produced power, which are carried out by methods based on ML. Due to the instability of this type of system, ML methods are used because the input of these systems are the weather conditions in the selected region. Due to their dependence on climate change, they have variable values and are not steady parameters. The utilized methods, like other extensive studies in estimating the produced power of RESs, have a reliable assessment of the stability. The most noticeable limitation of this study is the uncertainty of the VBT results. It is the most novel type of wind turbine, and its commercial versions are not used yet. So, the results are compared with the demo version. Furthermore, numerical solution and ML prediction are performed for comparison; in future studies, we will provide methods in order to estimate a produced power for the next few years as a climate change modeling.

Author Contributions: M.D.M., M.S. and A.M.—conceptualization; M.D.M., M.M. and M.S.—simulation; M.D.M., M.M. and A.M.—methodology; M.M. and A.M.—software; M.D.M., L.K., A.M. and M.S.—supervision; M.D.M., A.M. and L.K.—writing and review; M.M., M.S., A.M. and L.K.—administration; A.M.—funding acquisition. All authors have read and agreed to the published version of the manuscript.

Funding: The project no. 2019–1.3.1-KK-2019-00007 was implemented with support from the National Research, Development and Innovation Fund of Hungary, financed under the 2019–1.2.1-KK funding scheme. L.K. was supported in addition by the Eötvös Loránd Research Network Secretariat under grant agreement no. ELKH KÖ-40-2020 (Development of cyber-medical systems based on AI and hybrid cloud methods).

Data Availability Statement: The data is available through the corresponding author.

Acknowledgments: Amir Mosavi wishes to acknowledge the support of H2020, the European Union's Horizon 2020 Research and Innovation Programme under the Programme SASPRO 2 CO-FUND Marie Sklodowska-Curie grant agreement No. 945478.

Conflicts of Interest: The authors declare that they have no known competing financial interests or personal relationships that could have appeared to influence the work reported in this paper.

Abbreviations

Abbreviations	Description
ACC	Accuracy
AI	Artificial intelligence
ANN	Artificial neural network
CFD	Computational fluid dynamics
CO_2	Carbon dioxide
FPR	False positive rate
HBWTWEC	Hybrid site of the bladeless wind turbine and the wave energy converter
IOT	Internet of things
LSTM	Long-short term memory
MAE	Mean absolute error
ML	Machine learning
PPV	Positive predictive value
RESs	Renewable energy systems
RMSE	Root mean squared error
RNNs	Recurrent neural network
ROC	Receiver operating characteristic
SVM	Supported vector machine
TNR	True negative rate
TPR	True positive rate
VBT	Vortex bladeless turbine

References

1. Chakraborty, S.; Dwivedi, P.; Gupta, R.; Das, S. An Overview of Ocean Energy Policies Across the World. *Water Energy Int.* **2021**, *64*, 38–46.
2. Zandalinas, S.I.; Fritschi, F.B.; Mittler, R. Global warming, climate change, and environmental pollution: Recipe for a multifactorial stress combination disaster. *Trends Plant Sci.* **2021**, *26*, 588–599. [CrossRef] [PubMed]
3. Dincer, I.; Rosen, M.A. *Exergy: Energy, Environment and Sustainable Development*, 3rd ed.; Elsevier: Oxford, UK, 2021.
4. Olabi, A.G.; Abdelkareem, M.A. Renewable energy and climate change. *Renew. Sustain. Energy Rev.* **2022**, *158*, 112111. [CrossRef]
5. Gernaat, D.E.; de Boer, H.S.; Daioglou, V.; Yalew, S.G.; Müller, C.; van Vuuren, D.P. Climate change impacts on renewable energy supply. *Nat. Clim. Change* **2021**, *11*, 119–125. [CrossRef]
6. Fell, H.; Gilbert, A.; Jenkins, J.D.; Mildenberger, M. Nuclear power and renewable energy are both associated with national decarbonization. *Nat. Energy* **2022**, *7*, 25–29. [CrossRef]
7. Rony, J.S.; Karmakar, D. Coupled Dynamic Analysis of Hybrid Offshore Wind Turbine and Wave Energy Converter. *J. Offshore Mech. Arct. Eng.* **2022**, *144*, 032002. [CrossRef]
8. Si, Y.; Chen, Z.; Zeng, W.; Sun, J.; Zhang, D.; Ma, X.; Qian, P. The influence of power-take-off control on the dynamic response and power output of combined semi-submersible floating wind turbine and point-absorber wave energy converters. *Ocean Eng.* **2021**, *227*, 108835. [CrossRef]
9. Skene, D.M.; Sergiienko, N.; Ding, B.; Cazzolato, B. The Prospect of Combining a Point Absorber Wave Energy Converter with a Floating Offshore Wind Turbine. *Energies* **2021**, *14*, 7385. [CrossRef]
10. Cheng, Z.; Wen, T.R.; Ong, M.C.; Wang, K. Power performance and dynamic responses of a combined floating vertical axis wind turbine and wave energy converter concept. *Energy* **2019**, *171*, 190–204. [CrossRef]
11. Dong, C.; Huang, G.G.; Cheng, G. Offshore wind can power Canada. *Energy* **2021**, *236*, 121422. [CrossRef]
12. Dehghani-Sanij, A.R.; Al-Haq, A.; Bastian, J.; Luehr, G.; Nathwani, J.; Dusseault, M.B.; Leonenko, Y. Assessment of current developments and future prospects of wind energy in Canada. *Sustain. Energy Technol. Assess.* **2022**, *50*, 101819. [CrossRef]
13. Robertson, B.; Dunkle, G.; Gadasi, J.; Garcia-Medina, G.; Yang, Z. Holistic marine energy resource assessments: A wave and offshore wind perspective of metocean conditions. *Renew. Energy* **2021**, *170*, 286–301. [CrossRef]
14. Sadorsky, P. Wind energy for sustainable development: Driving factors and future outlook. *J. Clean. Prod.* **2021**, *289*, 125779. [CrossRef]

15. Latif, A.; Hussain, S.S.; Das, D.C.; Ustun, T.S. Double stage controller optimization for load frequency stabilization in hybrid wind-ocean wave energy based maritime microgrid system. *Appl. Energy* **2021**, *282*, 116171. [CrossRef]
16. Dawoud, S.M. Developing different hybrid renewable sources of residential loads as a reliable method to realize energy sustainability. *Alex. Eng. J.* **2021**, *60*, 2435–2445. [CrossRef]
17. Ammari, C.; Belatrache, D.; Touhami, B.; Makhloufi, S. Sizing, optimization, control and energy management of hybrid renewable energy system—A review. *Energy Built Environ.* **2021**, *3*, 399–411. [CrossRef]
18. Luderer, G.; Madeddu, S.; Merfort, L.; Ueckerdt, F.; Pehl, M.; Pietzcker, R.; Rottoli, M.; Schreyer, F.; Bauer, N.; Baumstark, L.; et al. Impact of declining renewable energy costs on electrification in low-emission scenarios. *Nat. Energy* **2022**, *7*, 32–42. [CrossRef]
19. Curto, D.; Franzitta, V.; Guercio, A. Sea Wave Energy. A Review of the Current Technologies and Perspectives. *Energies* **2021**, *14*, 6604. [CrossRef]
20. Rodrigues, C.; Ramos, M.; Esteves, R.; Correia, J.; Clemente, D.; Gonçalves, F.; Mathias, N.; Gomes, M.; Silva, J.; Duarte, C.; et al. Integrated study of triboelectric nanogenerator for ocean wave energy harvesting: Performance assessment in realistic sea conditions. *Nano Energy* **2021**, *84*, 105890. [CrossRef]
21. Choupin, O.; Andutta, F.P.; Etemad-Shahidi, A.; Tomlinson, R. A decision-making process for wave energy converter and location pairing. *Renew. Sustain. Energy Rev.* **2021**, *147*, 111225. [CrossRef]
22. Band, S.S.; Ardabili, S.; Mosavi, A.; Jun, C.; Khoshkam, H.; Moslehpour, M. Feasibility of soft computing techniques for estimating the long-term mean monthly wind speed. *Energy Rep.* **2022**, *8*, 638–648. [CrossRef]
23. Luo, H.; Cao, S.; Lu, V.L. The techno-economic feasibility of a coastal zero-energy hotel building supported by the hybrid wind-wave energy system. *Sustain. Energy Grids Netw.* **2022**, *30*, 100650. [CrossRef]
24. Kamarlouei, M.; Gaspar, J.F.; Calvario, M.; Hallak, T.S.; Mendes, M.J.; Thiebaut, F.; Soares, C.G. Experimental study of wave energy converter arrays adapted to a semi-submersible wind platform. *Renew. Energy* **2022**, *188*, 145–163. [CrossRef]
25. Wang, B.; Deng, Z.; Zhang, B. Simulation of a novel wind–wave hybrid power generation system with hydraulic transmission. *Energy* **2022**, *238*, 121833. [CrossRef]
26. Chennaif, M.; Maaouane, M.; Zahboune, H.; Elhafyani, M.; Zouggar, S. Tri-objective techno-economic sizing optimization of Off-grid and On-grid renewable energy systems using Electric system Cascade Extended analysis and system Advisor Model. *Appl. Energy* **2022**, *305*, 117844. [CrossRef]
27. Akorede, M.F. Design and performance analysis of off-grid hybrid renewable energy systems. In *Hybrid Technologies for Power Generation*; Academic Press: Cambridge, MA, USA, 2022; pp. 35–68.
28. Pravin, P.S.; Luo, Z.; Li, L.; Wang, X. Learning-based Scheduling of Industrial Hybrid Renewable Energy Systems. *Comput. Chem. Eng.* **2022**, *159*, 107665. [CrossRef]
29. Jahangir, M.H.; Shahsavari, A.; Rad, M.A. Feasibility study of a zero emission PV/Wind turbine/Wave energy converter hybrid system for stand-alone power supply: A case study. *J. Clean. Prod.* **2020**, *262*, 121250. [CrossRef]
30. Zhou, Y.; Ning, D.; Shi, W.; Johanning, L.; Liang, D. Hydrodynamic investigation on an OWC wave energy converter integrated into an offshore wind turbine monopile. *Coast. Eng.* **2020**, *162*, 103731. [CrossRef]
31. Aazami, R.; Heydari, O.; Tavoosi, J.; Shirkhani, M.; Mohammadzadeh, A. Optimal Control of an Energy-Storage System in a Microgrid for Reducing Wind-Power Fluctuations. *Sustainability* **2022**, *14*, 6183. [CrossRef]
32. Weerakoon, A.S.; Kim, B.H.; Cho, Y.J.; Prasad, D.D.; Ahmed, M.R.; Lee, Y.H. Design optimization of a novel vertical augmentation channel housing a cross-flow turbine and performance evaluation as a wave energy converter. *Renew. Energy* **2021**, *180*, 1300–1314. [CrossRef]
33. Chaurasia, K.; Kamath, H.R. Artificial Intelligence and Machine Learning Based: Advances in Demand-Side Response of Renewable Energy-Integrated Smart Grid. In *Smart Systems: Innovations in Computing*; Springer: Singapore, 2022; pp. 195–207.
34. Musbah, H.; Ali, G.; Aly, H.H.; Little, T.A. Energy management using multi-criteria decision making and machine learning classification algorithms for intelligent system. *Electr. Power Syst. Res.* **2022**, *203*, 107645. [CrossRef]
35. Hernandez, D.; Denis, Y. Energy Management System Industrialization for Off-Grids Power Systems Based on Data-Driven Machine Learning Models. Available online: https://ssrn.com/abstract=4003926 (accessed on 17 June 2021).
36. Patel, A.; Swathika, O.V.; Subramaniam, U.; Babu, T.S.; Tripathi, A.; Nag, S.; Karthick, A.; Muhibbullah, M. A Practical Approach for Predicting Power in a Small-Scale Off-Grid Photovoltaic System using Machine Learning Algorithms. *Int. J. Photoenergy* **2022**, *2022*, 9194537. [CrossRef]
37. Abualigah, L.; Zitar, R.A.; Almotairi, K.H.; Hussein, A.M.; Abd Elaziz, M.; Nikoo, M.R.; Gandomi, A.H. Wind, Solar, and Photovoltaic Renewable Energy Systems with and without Energy Storage Optimization: A Survey of Advanced Machine Learning and Deep Learning Techniques. *Energies* **2022**, *15*, 578. [CrossRef]
38. Buster, G.; Bannister, M.; Habte, A.; Hettinger, D.; Maclaurin, G.; Rossol, M.; Sengupta, M.; Xie, Y. Physics-guided machine learning for improved accuracy of the National Solar Radiation Database. *Sol. Energy* **2022**, *232*, 483–492. [CrossRef]
39. Zou, S.; Zhou, X.; Khan, I.; Weaver, W.W.; Rahman, S. Optimization of the electricity generation of a wave energy converter using deep reinforcement learning. *Ocean Eng.* **2022**, *244*, 110363. [CrossRef]
40. Al-Othman, A.; Tawalbeh, M.; Martis, R.; Dhou, S.; Orhan, M.; Qasim, M.; Olabi, A.G. Artificial intelligence and numerical models in hybrid renewable energy systems with fuel cells: Advances and prospects. *Energy Convers. Manag.* **2022**, *253*, 115154. [CrossRef]

41. Mousavi, S.M.; Ghasemi, M.; Dehghan Manshadi, M.; Mosavi, A. Deep learning for wave energy converter modeling using long short-term memory. *Mathematics* **2021**, *9*, 871. [CrossRef]
42. Dehghan Manshadi, M.; Ghassemi, M.; Mousavi, S.M.; Mosavi, A.H.; Kovacs, L. Predicting the Parameters of Vortex Bladeless Wind Turbine Using Deep Learning Method of Long Short-Term Memory. *Energies* **2021**, *14*, 4867. [CrossRef]
43. He, J. Coherence and cross-spectral density matrix analysis of random wind and wave in deep water. *Ocean Eng.* **2020**, *197*, 106930. [CrossRef]
44. Eltohamy, I. Effect of Vertical Screen on Energy Dissipation and Water Surface Profile Using Flow 3D. *Egypt. Int. J. Eng. Sci. Technol.* **2022**, *38*, 20–25.
45. Narasimhan, A. Support Vector Machine Based Forecasting for Renewable Energy Systems. In *Renewable Energy Optimization, Planning and Control 2022*; Springer: Singapore, 2022; pp. 149–157.
46. Babajani, A. Hydrodynamic performance of a novel ocean wave energy converter. *Am. J. Fluid Dyn.* **2018**, *8*, 73–83.
47. Farah, S.; Humaira, N.; Aneela, Z.; Steffen, E. Short-term multi-hour ahead country-wide wind power prediction for Germany using gated recurrent unit deep learning. *Renew. Sustain. Energy Rev.* **2022**, *167*, 112700. [CrossRef]
48. Neshat, M.; Nezhad, M.M.; Abbasnejad, E.; Mirjalili, S.; Groppi, D.; Heydari, A.; Wagner, M. Wind turbine power output prediction using a new hybrid neuro-evolutionary method. *Energy* **2021**, *229*, 120617. [CrossRef]
49. Lu, P.; Ye, L.; Zhao, Y.; Dai, B.; Pei, M.; Tang, Y. Review of meta-heuristic algorithms for wind power prediction: Methodologies, applications and challenges. *Appl. Energy* **2021**, *301*, 117446. [CrossRef]

Review

Site Selection of Combined Offshore Wind and Wave Energy Farms: A Systematic Review

Shabnam Hosseinzadeh [1,2], Amir Etemad-Shahidi [1,2,3,*] and Rodney A. Stewart [1,2]

1. School of Engineering and Built Environment, Griffith University, Southport, QLD 4222, Australia
2. Cities Research Institute, Griffith University, Southport, QLD 4222, Australia
3. School of Engineering, Edith Cowan University, Joondalup, WA 6027, Australia
* Correspondence: a.etemadshahidi@griffith.edu.au

Abstract: Growing energy demand worldwide and onshore limitations have increased interest in offshore renewable energy exploitation. A combination of offshore renewable energy resources such as wind and wave energy can produce stable power output at a lower cost compared to a single energy source. Consequently, identifying the best locations for constructing combined offshore renewable energy farms is crucial. This paper investigates the technical, economic, social, and environmental aspects of Combined Offshore Wind and Wave Energy Farm (COWWEF) site selection. Past literature was evaluated using a systematic review method to synthesize, criticize, and categorize study regions, dataset characteristics, constraints, evaluation criteria, and methods used for the site selection procedure. The results showed that most studied regions belong to European countries, and numerical model outputs were mainly used in the literature as met-ocean data due to the limited coverage and low spatiotemporal resolution of buoy and satellite observations. Environmental and marine usage are the main constraints in the site selection process. Among all constraints, shipping lanes, marine protected areas, and military exercise areas were predominately considered to be excluded from the potential sites for COWWEF development. The technical viability and economic feasibility of project deployment are emphasized in the literature. Resource assessment and distance to infrastructures were mostly evaluated among techno-economic criteria. Wind and wave energy power are the most important criteria for evaluating feasibility, followed by water depth, indicators of variability and correlation of the energy resources, and distance to the nearest port. Multi-Criteria Decision-Making (MCDM) methods and resource-based analysis were the most-used evaluation frameworks. Resource-based studies mainly used met-ocean datasets to determine site technical and operational performance (i.e., resource availability, variability, and correlation), while MCDM methods were applied when a broader set of criteria were evaluated. Based on the conducted review, it was found that the literature lacks evaluation of seabed conditions (seabed type and slope) and consideration of uncertainty involved in the COWWEF site selection process. In addition, the market analysis and evaluation of environmental impacts of COWWEF development, as well as impacts of climate change on combined exploitation of offshore wind and wave energy, have rarely been investigated and need to be considered in future studies. Finally, by providing a comprehensive repository of synthesized and categorized information and research gaps, this study represents a road map for decision-makers to determine the most suitable locations for COWWEF developments.

Keywords: offshore wind energy; wave energy; site selection; multi-criteria decision-making; resource assessment; restrictions; evaluation criteria

Citation: Hosseinzadeh, S.; Etemad-Shahidi, A.; Stewart, R.A. Site Selection of Combined Offshore Wind and Wave Energy Farms: A Systematic Review. *Energies* **2023**, *16*, 2074. https://doi.org/10.3390/en16042074

Academic Editors: Yan Bao, Guanghua He and Liang Sun

Received: 16 January 2023
Revised: 13 February 2023
Accepted: 18 February 2023
Published: 20 February 2023

Copyright: © 2023 by the authors. Licensee MDPI, Basel, Switzerland. This article is an open access article distributed under the terms and conditions of the Creative Commons Attribution (CC BY) license (https://creativecommons.org/licenses/by/4.0/).

1. Introduction

Presently, population growth and industrialization have led to a rise in energy demand across the globe. Traditional fossil energy resources are limited and will be depleted in the future. In addition, burning fossil fuels has caused air pollution and raised environmental

concerns. The Paris Agreement was adopted to reduce emissions to net zero by 2050. The international community also completed the rulebook of this agreement by signing the Glasgow Climate Pact. A decade of climate action and support is therefore an aim for the 2020s [1]. For these reasons, the trend toward using clean and sustainable alternatives, namely renewable energy resources, has increased [2].

Renewable energy resources' share of the world's total energy consumption was approximately 13% in 2020 [3]. Biomass, geothermal, solar, hydropower, wind, and marine are renewable energy resources, the environmental impacts of which are insignificant compared to conventional energy sources and, therefore, are appropriate to supply the future clean energy demand [4]. The global renewable energy generation's largest share at the end of 2020 was hydropower (57%), followed by wind energy (21%), solar power (11%), and others (11%). Among renewables, wind and solar energy's contributions to the world energy capacity have been significant [5].

In recent years, offshore renewable power plant developments have become popular due to their power capacity and generation potential as well as limited onshore space. Several countries are progressively undertaking new projects regarding offshore wind power plant development [6]. The abundance of available space, low noise, and less visual impact encourage planners to develop offshore wind energy farms with fewer constraints on wind turbine (WT) size and less environmental impact [7]. The wind on open seas supplies a high level of power generation [8]. The reduced turbulence in offshore areas because of lower surface roughness, compared to onshore wind, provides higher offshore wind speed. Significant energy potential is thus expected in offshore areas, as the power is proportional to cubic wind speed [9]. However, the primary concern in offshore wind farms is their accessibility during unsuitable weather conditions [10].

Ocean wave energy is a widely untapped renewable energy source and has the potential to influence worldwide energy production [11]. The wave energy resource is power-dense and continuous and thus reliable for energy production [11,12]. Higher wave energy density signifies more energy extraction from a smaller ocean volume at a lower cost [13]. Low visibility and marine environment protection by attracting wave energy are other advantages of Wave Energy Converters (WECs). Nevertheless, the WECs' development has been slow due to technical issues and economic obstacles [14]. The survivability in extreme conditions and optimization of efficiency are existing challenges in exploiting wave energy [15,16].

1.1. Combined Offshore Wind and Wave Energy Farms

Due to increasing levels of investment in offshore renewable energy systems, enhanced site selection methods are required to minimize the costs. Offshore renewable energy resource (wind and wave) variations, which come from their characteristics of randomness and intermittency, adversely affect energy generators' efficiency and hence lead to higher energy costs [17]. The synergy of renewable energy resources is an efficient solution to resolve challenges presented by standalone renewables and optimize energy exploitation. Coupling complementary renewable energy options ensures greater reliability of energy supply by reducing the variation of power output and downtime period [17,18]. Integrating offshore wind and wave energy extraction makes the energy output more reliable and higher than the sum of disconnected farms [17]. Sharing space, grid connection, and infrastructures (e.g., foundations) in coupled wind and wave energy farms reduces construction and maintenance costs and improves efficiency [19]. In addition, joint exploitation of wind and wave energy resources reduces the structural load and makes it easier to access offshore wind power systems [20,21].

1.2. Site Selection Process

The crucial step prior to renewable energy farm development is site selection. To ensure the success of combined offshore wind and wave energy projects, selecting the most appropriate location for the plants' installation makes power production more favorable from technical, economic, social, and environmental perspectives. Various constraints and objectives must be considered for optimal and practical renewable energy farms. The main concern is reducing costs while having the highest efficiency [22,23]. Other marine usage areas, such as fisheries and shipping routes, should be considered to reduce environmental impacts. In order to identify sustainable siting, many conflicting criteria and aspects of technical constraints, economic feasibility, and environmental and social impacts need to be evaluated [24]. Hence, a transparent and reliable framework is required to integrate these conflicting factors and make a final decision [25]. These inconsistent criteria make site selection a Multi-Criteria Decision-Making (MCDM) problem.

1.3. Existing Literature and Purpose of the Study

The literature is devoted to trends in renewable energy studies and technological advancements. A systematic review [26] highlighted the key criteria for assessing the feasibility of offshore wind energy deployment. In [27], all factors to optimize onshore and offshore wind power locations were categorized, underlining the differences between their decision criteria. The evaluation of trends in offshore wind energy research by [28] highlighted the use of GIS as a common site selection tool. In [29], site selection procedures in both onshore and offshore wind energy were analyzed. The application of MCDM methods for site selection of renewable energy resources was reviewed in [30], and the used Exclusion (EX) and Evaluation (EV) criteria were summarized. In [31], the restrictive and deterministic factors and methodologies for the site selection of onshore wind power plants were assessed.

As noted, most review papers examine standalone renewable energy farms' site selections. In the case of combined offshore renewable energy exploitation, only the technological aspect of energy systems is investigated in the literature. For example, a review of synergetic technologies capable of hybridization with wave energy was conducted by [32]. Moreover, the structural options and technological aspects of combined offshore wind and wave energy systems were reviewed by [33]. In contrast with the existing abovementioned review papers, the main novelty of this paper is comprehensively reviewing the site selection process for combined offshore wind and wave energy exploitation. In this review, different perspectives of COWWEF deployment (technical, economic, social, and environmental) in the context of site selection are discussed to efficiently reduce its associated cost and negative environmental and social impacts while maximizing energy production efficiency. Regarding the growing number of studies in this field, having an overview of performed studies draws a roadmap for the future. In this way, the studied regions, the met-ocean dataset characteristics, the exclusion and evaluation criteria, and the applied methodologies for site selection analysis of COWWEF are categorized and synthesized in this paper. Practitioners and decision-makers can use the comprehensive information provided in this study to identify the optimal sites for the installation of joint wind–wave power plants. The following research questions summarizing the most critical aspects of the selection of suitable locations for the COWWEF are addressed in this paper:

1. Where are the studied regions?
2. Which types of met-ocean datasets were employed (i.e., observational or modeled, resolution, and duration)?
3. What exclusion criteria restrict the selection process?
4. What evaluation criteria influence the determination of hotspots?
5. Which methodologies were used for site selection?

2. Methods

In order to answer the questions outlined in the previous section, this study was conducted according to the guidelines of the Systematic Literature Review (SLR). Systematic reviews seek objective and impartial responses to specific questions [34]. For this purpose, a systematic method defined as a priori in studies identification and selection, data extraction, and result analysis was employed. In the systematic review process, it is necessary to map the review's objectives and the process of finding studies prior to proceeding with the systematic review [35]. A search strategy was therefore developed to systematically find articles following the guidelines used in the literature [31]. Figure 1 illustrates the main steps taken to obtain the results. As can be seen from the figure, this Systematic Literature Review (SLR) includes three steps: identifying relevant papers, excluding irrelevant and duplicated studies, and checking the publications' eligibility and inclusions. The following subsections describe these steps.

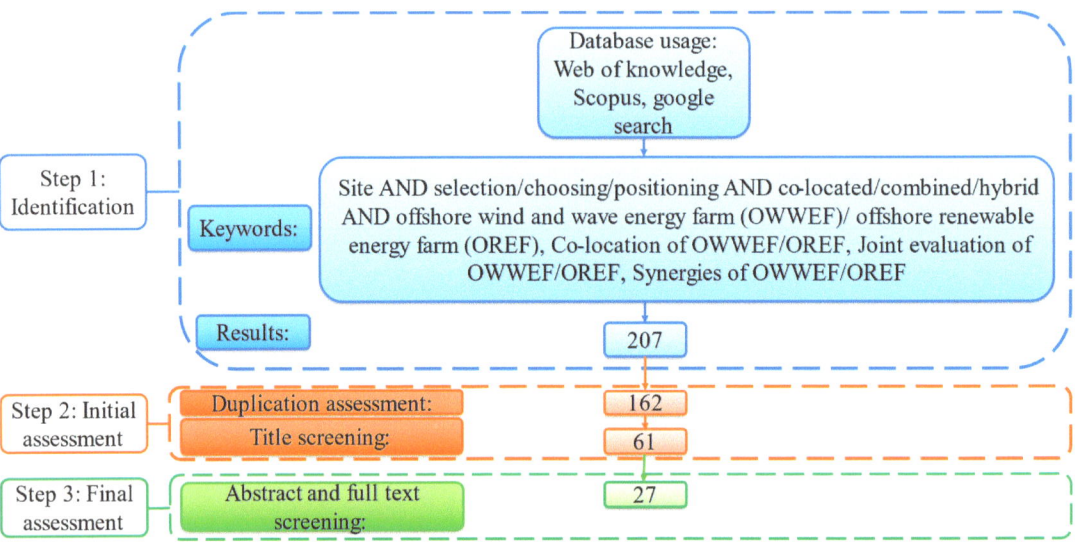

Figure 1. Flowchart for selection of existing studies based on systematic literature review guidelines.

2.1. Step 1: Identification

First, the identification of records was performed through the Web of Knowledge database, Scopus, and Google searches. Choosing the Web of Knowledge and Scopus search platforms was based on the wide range of high-quality journals available there, which raise the quality of the results obtained for this study. Google searches can also be helpful for finding relevant project reports. The keywords were selected based on the review purpose and the most common keywords used in the topic area. A series of pilot searches were conducted by trial and error to refine the keywords used in the search string. Terms that did not contribute any additional results to the automatic search were removed. Based on the search results, the existing publications were found from 2009 to 2022.

2.2. Step 2: Exclusion of Irrelevant and Duplicated Studies (Initial Assessment)

According to the obtained results from the first step, 207 publications were found. As the first stage of the initial assessment, removing duplicate studies using the Mendeley reference manager resulted in 162 documents. The remaining publications were reviewed through a practical screening method, including establishing exclusion and inclusion criteria. For separating the publications that contributed information and answered the research questions considered in this study from those that did not, the exclusion criteria

were applied. Therefore, the title of each paper was checked considering the following exclusion criteria: Is the publication about offshore renewable energy (wind and wave)? Is the site selection issue or resource assessment discussed? If not, the publication was excluded. This process yielded 61 publications, as most did not properly fit the keyword-assigned theme.

2.3. Step 3: Checking the Eligibility of Publications (Final Assessment)

The final assessment was conducted to verify the eligibility of the remaining publications. This phase was conducted based on reading the abstract and screening the full text of the articles to find answers to the research questions. The inclusion criteria were as follows: Does the publication discuss COWWEF site selection? Have the factors for site selection been mentioned? Does the publication include a methodology for the optimal positioning of COWWEF? Eventually, 27 relevant publications were found and considered for review in this paper.

The final selected studies were analyzed to answer the research questions, and a spreadsheet was created to categorize studies in terms of study regions, met-ocean data specifications, EX and EV criteria, and methodologies.

3. Results and Discussion

3.1. The Studied Regions

Figure 2 illustrates the geographic locations of selected papers for positioning COWWEFs. European offshore areas were assessed for site selection purposes in most of the studies (22), while a few studies included other continents [36–41]. This is because of the European Union's long-term strategies to increase energy security [42]. In addition, aesthetic concerns and a lack of available shallow-water locations in European areas have driven the need to deploy offshore WTs. These areas are known for stronger and more consistent prevailing winds [43]. Certain technical obstacles in the wave energy industry must be eliminated to achieve energy efficiency, cost-effectiveness, safety, and durability [42]. The aforementioned reasons led to trends in developing COWWEFs, to simultaneously exploit energy resources. This can reduce the associated cost by enhancing the energy yield, producing reliable energy, and using shared grid infrastructures [33].

3.2. Used Met-Ocean Dataset for Site Selection

Buoy and satellite data are sparse and only available for a limited number of periods [44]. A significant advantage of satellite measurements is coverage of large areas, whereas buoys are only capable of sampling in a single location. On the other hand, buoys can take measurements constantly with a high temporal resolution. In contrast, many ocean-sensing instruments placed on polar-orbiting satellites cannot continuously monitor the same location [45]. In order to provide data in the absence of these long-term observations with high temporal and spatial resolution, numerical models have mainly (93%) been used in the literature. Table 1 displays the details of the used dataset (mainly ECWMF) in the literature. The temporal resolutions of data vary among 1, 3, and 6 h and daily time scales. However, they mostly have hourly resolutions. The spatial resolution used in existing studies varies from 17 m to 83.25 km.

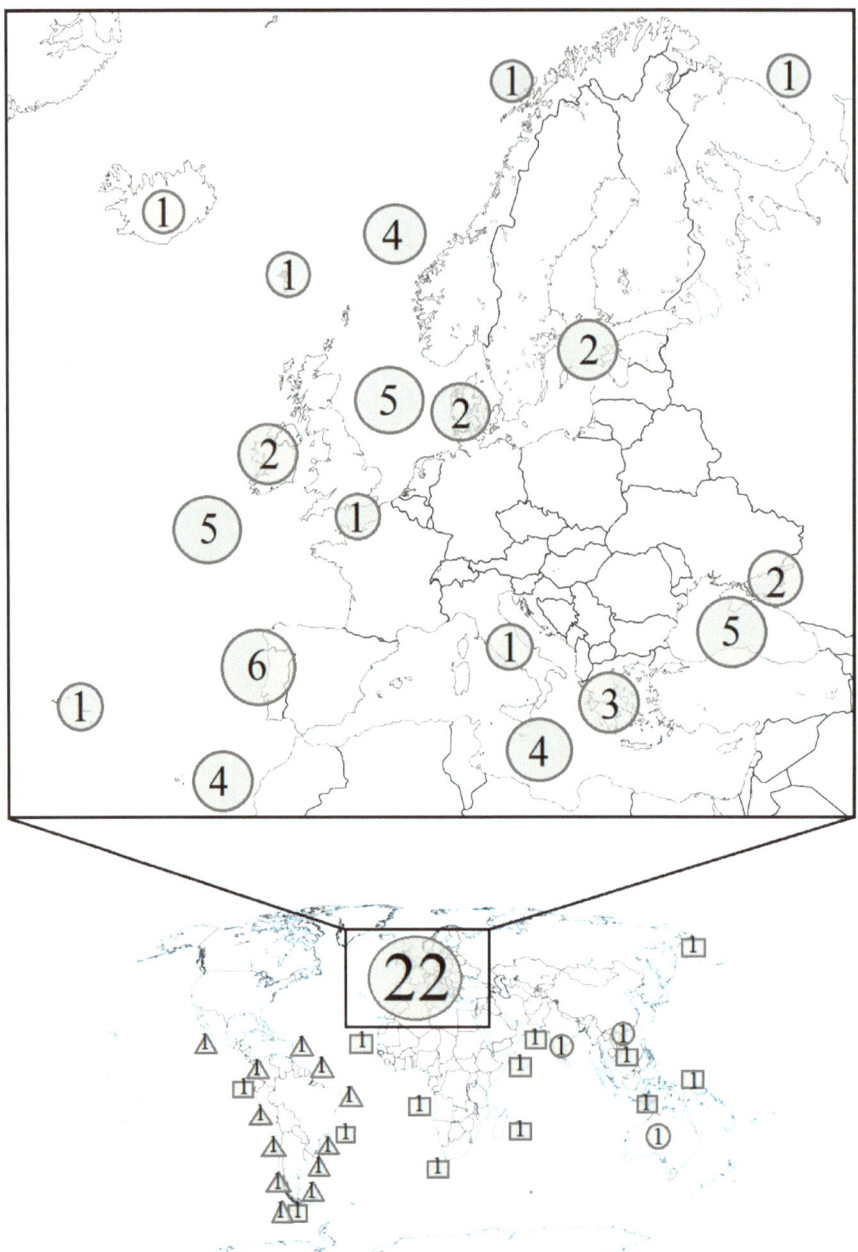

Figure 2. Geographic locations of existing studies for combined offshore wind and wave energy site selection (the areas represented in the rectangle and the triangle are related to two studies with considerable numbers of studied sites around the world). Note that a study performed for all offshore areas around the world [41] is not shown in this figure.

Table 1. Used met-ocean dataset details.

Dataset	Duration	Resolution (Temporal and Spatial)	Reference
Observational data from buoys	Jan 2002–Jan 2005	Hourly	[17]
ECMWF 40 (ERA-40) Global wave forecast dataset	1987–2009	Wind: 3 h, Wave: 6 h The data were interpolated in an area divided into 300 cells of 22×12 km of grid size.	[46]
WAM and REMO	Jan 1958–Dec 2001	3 h WAM: 9.25 km × 9.25 km REMO: 55.5 km × 55.5 km	[47]
WAM and REMO	Jan 1958–Dec 2001	3 h WAM: 9.25 km × 9.25 km REMO: 55.5 km × 55.5 km	[20]
ECMWF version of WAM and SKIRON atmospheric model	2001–2010	Hourly, 5.55 km × 5.55 km	[19]
WAsP and SWAN	2005–2015	Hourly An area of 134 km × 167 km with a resolution of 300 m × 300 m and a nested grid covering 8.5 km × 8.5 km with a resolution of 17 m × 17 m	[48]
CFSR and SWAN	1997–2016	3 h Wind: 37.05 km × 37.05 km Wave: 0.888 km × 0.888 km	[49]
ERA	2001–2016	6 h, 83.25 km × 83.25 km	[37]
SWAN and WAsP	Feb 2005–Jan 2015	Hourly, 2.775 km × 2.775 km	[50]
ERA	Jan 2005–Dec 2014	6 h, 83.25 km × 83.25 km	[51]
SKIRON-Eta and WAM	1995–2004	3 h, 11.1 km × 11.1 km	[37]
SKIRON and WAM	2001–2010	Hourly, 5 km × 5 km	[52]
NCEP CFSR and GOW2	1979–2015	Hourly Wind: 33.3 km × 33.3 km in the period of 1979–2010 and 22.2 km × 22.2 km in the period of 2011–2015 Wave: 27.75 km × 27.75 km	[41]
NCEP-CFSR and SWAN	Wind: 1997–2016 Wave: 1999–2013	3 h Wind: 34.63 km × 34.63 km Wave: 8.88 km × 8.88 km	[53]

Table 1. *Cont.*

Dataset	Duration	Resolution (Temporal and Spatial)	Reference
SKIRON-Eta and WAM	1995–2004	3 h, 11.1 km × 11.1 km	[54]
NCEP-CFCR and SWAN	1987–2016	Wind: 34.6 km × 34.6 km Wave: 148 m and 37 m	[55]
ECMWF ERA-Interim	2005–2014	Wind: 3 h Wave: 6 h 60 km × 50 km	[56]
ECMWF ERA-Interim	Jan 2000–Dec 2016	6 h, 82 km × 82 km	[36]
Wind: WRF 3.3.1 model forced by CFSR Wave: WaveWatchIII forced by WRF	1979–2016	Hourly, 10 km × 10 km	[57]
Observational data from buoys	2009–2019	Hourly	[58]
Wind: Reginal climate simulations within CORDEX project under the RCP8.5 Wave: Dynamically downscaled of SWAN simulations forced by MIROC5GCM	2026–2045	Daily, 12.21 km × 12.21 km	[59]
Hindcast wind: WRF forced by CFSR Hindcast wave: WaveWatchIII Forecast wind and wave: Ensemble of nine models of GCM-RCM provided by EURO-CORDEX under the climate change scenario of RCP8.5	Hindcast wind: Jan 1979–Dec 2020 Forecast: 2030–2060 and 2070–2100	Hindcast of wind and wave: Hourly, 10 km Forecast wind: 6 h, 12.5 km Forecast wave: 3 h, 14.1 km × 9.99 km	[60]
ERA5 produced by ECMWF	Jan 2000–Dec 2019	Hourly Wave: 55.5 km × 55.5 km Wind: 27.75 km × 27.75 km	[61]
ECMWF	2016–2020	13.88 km × 13.88 km	[40]
Wind: NASA QuikSCAT satellite measurements Wave: Furgo-OCEANOR wave data originated by ECMWF WAM	Wind: 1999–2010	Hourly, 27.75 km × 27.75 km	[62]
ERA5 the latest global 150 atmospheric reanalysis product from ECMWF WaveWatchIII	2000–2018	Wind: Hourly, 31 km Wave: 6 h, 11.1 km × 11.1 km	[39]
Wave: WaveWatchIII model undertaken by CAWCR Wind: CFSR	2014–2020	Hourly, 7 km × 7 km	[38]

3.3. Restricted Areas for Site Selection

Some restrictions prevent farms from operating in the desired location. These may exclude or reduce the suitability of certain areas. Considering these restrictions could therefore result in a competition for space between new development and current users [51]. Investigation of existing studies revealed 12 restricted areas for offshore renewable energy developments in two categories: environmental and marine (i.e., sea and subsea) usage restrictions. The restricted areas were excluded in most studies. However, in a few studies, some areas were evaluated and assigned a low score in the site selection process (e.g., [46,51,57]). The two types of considerations are Exclusion (EX) and Evaluation (EV). Figure 3 outlines all restrictions in the existing studies, the consideration type in the literature (i.e., EX or EV), and their frequency of occurrence. The criteria values mentioned in the literature (e.g., minimum distance to the shore and shipping density) were noted in the text. Future studies could use these values to select the COWWEF location. However, planners should ensure that the laws of each country are followed to avoid potential disputes.

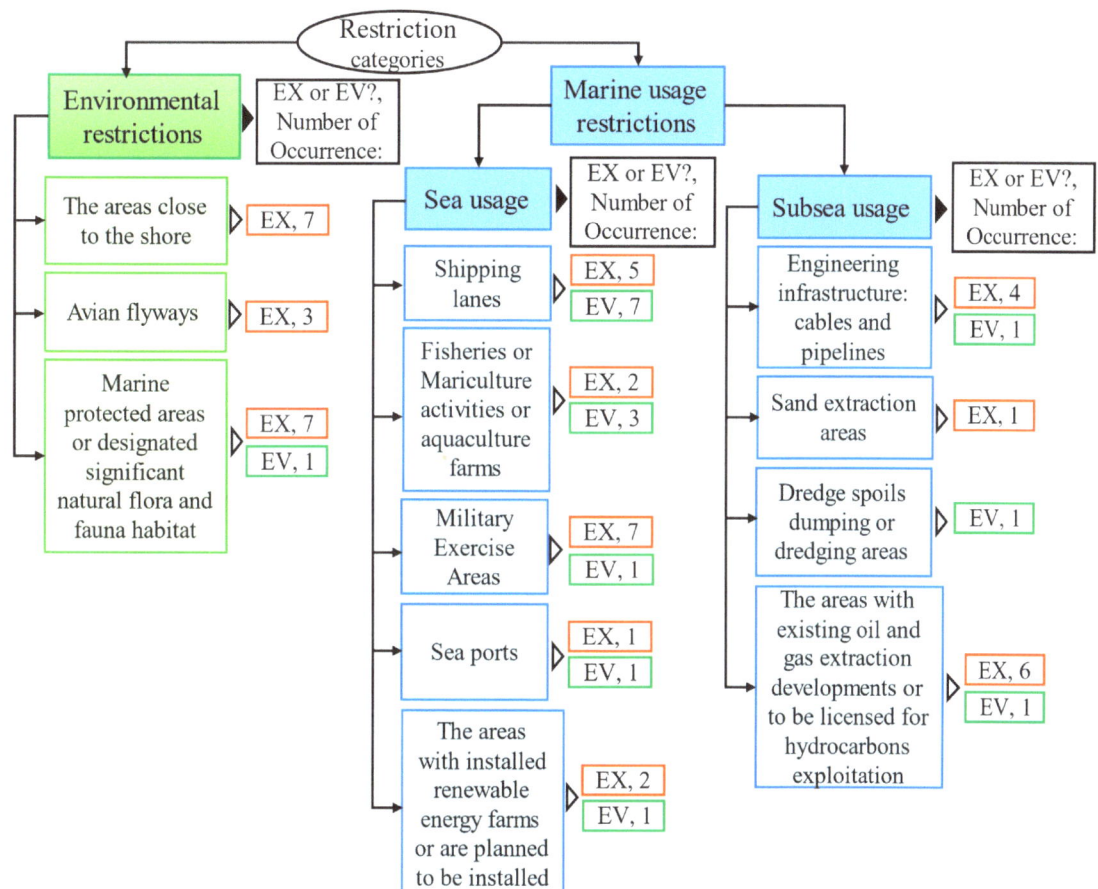

Figure 3. Graphical representation of restrictions (EX and EV refer to consideration type of criteria in literature, namely exclusion and evaluation, respectively).

Figure 3 shows that the main restriction is related to the usage of the sea and subsea. Among the 12 listed restricted areas, the most cited ones are the shipping lanes, military

exercise areas, marine protected areas, and areas close to the shore (due to visual and noise impact). In some studies, all shipping routes have been excluded from the analysis (e.g., [20,39,47]). However, in most studies (e.g., [42,46,54,56,62]), Shipping Density (SD) was used as an EV criterion. Defined as the number of ship tracks in a 1 km2 cell in [19], the SD was used as both EX and EV by [19,40]. In this way, high-traffic areas (e.g., SD > 25 and >75 [19] and >20 [40]) were excluded, and the shipping density in other areas was scored. In addition, giving the lowest weight to the SD by [62] allows the negotiability of shared use of an area.

Aquaculture farms and fisheries should also be considered. Using heavy towed fishing gear may damage transmit cables. Additionally, snagging fishing gear on cables could cause a ship to capsize or sink. Fishing vessels equipped with appropriate fishing gear may be allowed into COWWEFs if local fishers and offshore renewable energy operators reach an agreement [63]. In [50], the authors evaluated overlapping areas with fisheries and prioritized the region with a lower load of fisheries (see also [46,56]). Unlike the aforementioned studies, in which the lower load of fisheries was allowed to exist along with COWWEF, the coexistence of aquaculture farms with COWWEFs was considered unfeasible in [40]. In [62], the authors stated that fishing ports could be used for the operation and maintenance servicing of renewable energy devices but may not have a suitable draft for installation vessels. Areas with military exercise activities are also deemed unsuitable [42]. Most studies excluded these areas [40,42]. However, a low priority to these zones was assigned by [56] rather than excluding them. Sea ports were also excluded to avoid navigation interference [40]. The main port was given a low score due to the consideration of other marine users [56]. Furthermore, the areas planned to be installed or with installed renewable energy farms as existing sea usage were excluded in [42,54]. In contrast, these areas were given a low score by [56]. The construction of a power plant cannot be considered in the vicinity of existing subsea infrastructures such as cables and pipelines. Safety distances of 500 [50] and 920 m [39] were regarded as an exclusion zone. In [40], the cables and natural gas pipelines were excluded from the feasible areas (see also [62]). However, the areas with the above features were somehow evaluated and given a low score by [56]. Dredging zones and areas allocated to sand, oil, and gas extractions should not be selected, as they are already occupied. The need for access to the platforms should be considered when planning offshore renewable energy projects near oil and gas platforms [63]. The overlapping of these areas with the areas of high wind and wave energy resources was investigated by [50]. The sites to be licensed for the exploitation of hydrocarbons were considered ineligible for deploying COWWEFs in [42,54]. The oil and gas exploitation zones were removed from the eligible areas in some studies [39,40,62]. In addition, although dredging and oil and gas extraction regions were not excluded, a low preference was given to them by [56].

Mitigating the environmental impact of offshore renewable energy developments was also considered in the site selection. The areas close to the shore are considered environmentally restricted due to the visual and noise impact of offshore renewable energy farm developments. The literature thus proposed a minimum distance of 10 km [40,48], 15 km [19,39], 20 km [62], and 25 km [42,54] from shore. Bird corridors, marine protected areas, and natural flora and fauna habitats are highly sensitive. Some authors excluded avian flyways in the site selection process [20,47,50] due to the possibility of collisions of birds with WT blades. Marine-protected areas were eliminated from the eligible areas in most studies to protect the natural habitat [19,39,40,42,50,54,62]. In order to ensure safety, a 1 km buffer zone around marine protected areas was considered by [19]. In [56], natural marine habitats were given a low preference for renewable developments.

As can be seen from Figure 3, there is a consensus among most of the studies to completely exclude the abovementioned restricted areas other than shipping lanes, aquaculture farms, and dredging areas from eligible locations for COWWEF deployment. Considering the coexistence of lower loads of fisheries and shipping lanes with the energy farms can be related to the different policies of countries or the specific condition of the area which

dictates this consideration. In terms of dredging areas and some others (areas with cables and pipes, oil, and gas extraction platforms, installed devices, and marine protected areas), a low score was assigned by [56]. However, this quantification does not guarantee excluding those areas from the final proposed suitable locations. While the mentioned areas are required to exclude, there is no way to coexist with the areas planned for renewable energy farm development. Therefore, this is a disadvantage of the applied method in [56] for quantifying restricted areas.

3.4. Criteria for Evaluation in the Site Selection Procedure

In addition to the restrictions mentioned above, considering different perspectives of COWWEF site selection, including technical, economic, social, and environmental, using the relevant criteria, this study investigates the technical viability, economic feasibility, social acceptability, and environmental safety of the project deployment. Table 2 summarizes the criteria and their frequency of occurrence to evaluate the preferred siting of COWWEFs. More details, including the formula used for calculating some criteria and the acceptable range of criteria values, can be found in the Appendix A (Table A1). As can be seen from Table 2, the main objectives of consideration of the used criteria are: techno-economic analysis including (1) assessing the wind and wave climate data in terms of resources richness, variability, and complementarity, (2) considering the structural (device survivability) and technical (foundation/anchoring design of devices, installation, operation, and maintenance) feasibility of developments, (3) checking the accessibility of energy devices as an indicator of transmission cost and energy dissipation; economic analysis of feed-in tariff; socio-economic analysis with evaluation of the possibility of supplying energy demand; environmental analysis including (1) assessing the environmental impact of COWWEFs, and (2) examining the impact of human activities and environmental vulnerabilities in the site selection process. Therefore, the economic aspect of the project development is evaluated indirectly using three categories of criteria: techno-economic, economic, and socio-economic. However, the direct economic analysis is performed using the calculation of the Levelized Cost of Energy (LCOE), which is beyond the scope of this paper. In addition, based on Table A1 in the Appendix A, the lower and upper limits of the acceptable range for each criterion vary in different studies due to each studied region's local regulations, policy, and the considered energy-generation devices. For example, in [19], the areas with wind speed less than the minimum operational values of hypothetical devices (i.e., 6–7 m/s) were eliminated from the analysis (see also [42,54,62]). In terms of wave power density, a minimum value of 30 and 20 kw/m for the wave and wind dominated combined devices, respectively, was set by [19]. For other criteria, including water depth, distance from shore, port, local electrical grid, aquaculture, and nature conservation area, the locations with values out of the acceptable range were considered ineligible (see Table A1). The details of the used criteria in different categories of Tables 2 and A1, considering different aspects of COWWEF development, are discussed in the below subsections.

Table 2. Criteria used for evaluation, their relevant category, and frequency of mentions in the literature.

Category	Sub-Category	Criteria	No.
Techno-Economic	Wind energy resource richness	Wind Power, WP	20
		Wind Speed, WS	6
		Suitability Index of wind resource calculated based on percentage of time in which WP and Hs (significant wave height) are in acceptable range, $SI_{Wind\ R}$	1
		Total time with useful WS, $DWNT_{wind}$ (%)	1
		Rich level occurrence, RLO_{wind} (%)	1

Table 2. Cont.

Category	Sub-Category	Criteria	No.
	Wave energy resource richness	Wave Energy Power, WEP (kW/m)	25
		Significant wave height, Hs (m) and mean wave period, Tz (s)	2
		Suitability Index of wave resource calculated based on percentage of the time in which the WEP, Hs, and Tp are in the acceptable range, $SI_{wave\,R}$	1
		Total time with useful Hs, $DWNT_{Wave\,(\%)}$	1
	Combined offshore wind–wave farm richness	Mean Capacity Factor of combined energy farm, CF_{comb}	1
		Downtime or non-production time of combined energy farm, DT	1
	Resource variability	Coefficient of Variation, COV	11
		Monthly Variation, MV	5
		Seasonal Variation, SV	4
		Skewness, S	4
		Kurtosis, K	3
		Standard Deviation, SD	4
	Resource complementarity	Cross-Correlation Factor, CCF	11
		Complementarity indexes	1
	Devices' survivability	Suitability Index of structural survivability of wind device calculated based on the acceptable range of 50-year return period of significant wave height (H_{s50}), Wind Speed (WS_{50}) and current velocity (C_{50}), $SI_{Wind\,S}$	1
		50-year return period wind speed, WS_{50}	1
		Suitability Index of structural survivability of the wave device calculated based on the acceptable range of 50-year return period of significant wave height (H_{s50}) and current velocity (C_{50}), $SI_{Wind\,S}$	1
		50-year return period significant wave height, H_{s50}	1
	Foundation/anchoring design of devices	Water depth, WD	14
	Logistics (feasibility of installation, operation, and maintenance)	Suitability Index of logistics calculated based on the acceptable range of Distance from Port (DP), and the percentage of time in which WS and Hs are in the acceptable range, SI_{Log}	1
		Distance from Port, DP	9
		Distance from Shore, DS	9
	Transmission cost and energy dissipation	Distance to the Local Electrical Grid, DLEG	2
		Voltage Capacity of Closest available Grid, VCCG	3
Economic	Prioritizing different countries based on the feed-in tariff	Incentives: Feed-in tariffs of different countries located around the study region	1
Socio-Economic	Supplying energy demand	Population Served, PS	2
		Electricity Demand, ED	1
Environmental	Impact of offshore renewable energy farms on the environment	Environmental Performance Value, EPV	1

Table 2. *Cont.*

Category	Sub-Category	Criteria	No.
	Impact of considering human activities and environmental vulnerability in the site selection process	Cumulative Impact Index, CII	1
	Impact of noise on the growth of marine animals due to low frequency of sound waves	Distance from Aquaculture Area, DAA	1
	Impact of hitting birds by turbine blades	Distance from Nature Conservation Areas, DNCA	1

3.4.1. Techno-Economic Criteria

The most crucial aspect of site selection of renewable energy farms is the assessment of energy resource richness or availability. As shown in Tables 2 and A1, wind speed, wave height, and period are the main indicators used to calculate wind and wave energy power. As can be seen from the tables, wind and wave power are calculated in most studies to evaluate the energy resource's potential, while a few researchers directly used wind speed and wave parameters for the assessment of resource richness (e.g., [19,42,54,59,62]).

In [41], wave parameters and wind and wave energy power were used to assess the availability of energy resources. In this way, the percentage of time in which WP and Hs are in the acceptable ranges (WP $\geq$ 400 and Hs $\leq$ 5) was used to calculate the wind energy potential suitability index (SI Wind R). In the case of wave energy resource evaluation, a suitability index (SI Wave R) was defined based on the weighted average of the percentage of time in which wave energy power (WEP), significant height (Hs), and peak wave period (Tp) were in the acceptable ranges (WEP $\geq$ 15, $1 \leq$ Hs ≤ 6, and $5 \leq$ Tp ≤ 14).

The DNWT index was used by [59] to investigate the availability of energy resources considering the operational range of WTs and WECs. The typical cut-in and cut-off thresholds for WS are 4 and 25 ms^{-1}, respectively [59,64,65]. In [66], the upper and lower limits of 8 and 1 m for Hs were used, respectively, representing the power outage level due to extreme wave conditions and calm periods. In addition, the rich level occurrence (RLO) index was used by [59] to measure the frequency of wind power density higher than 200 Wm^{-2}. In [38], the mean capacity factor of different percentages of combined hypothetical WT and WEC (CF comb) was considered as an EV criterion to represent the efficiency of offshore renewable energy exploitation. In addition, the non-production time or Downtime (DT) of mixed offshore wind and wave energy farms (in which the energy devices are not operating) was considered to reflect the zero-power production amount [38].

Variability in wave and wind conditions plays a significant role in determining a location because peak-to-average ratios are a key cost factor [50]. The energy extraction devices are run in a specific range of wind and wave power so that their fluctuations and frequent on/off lower their efficiency leading to higher electricity costs [17]. Several statistical indicators were used to analyze the variability of energy resources: Standard Deviation (SD), Coefficient of Variation (COV), Total Harmonic Distortion (THD), Kurtosis (K) and Skewness (S), Monthly Variation (MV), and Seasonal Variation (SV) (see Table 2). The COV (i.e., the ratio of standard deviation to mean value) is the most cited index (37%) used in the literature. The THD with the same meaning has also been used to check the variability of offshore wind and wave energy resources separately and of mixed offshore wind and wave energy resources [36,37,51]. The THD for the hybrid offshore wind and wave energy farm is defined as the summation of wind and wave power's SDs divided by the summation of mean values of wind and wave. In addition, the COV index was obtained based on power production in the case of different percentages of using specific WT and WEC combinations by [38]. S and K are the other indicators used [39,52,53,55] to assess the variability of energy resources. S and K show how symmetric and heavy-tailed the data distribution is compared to a normal distribution. The variability of energy resources in the monthly and seasonal scales has also been evaluated in some existing studies using indices

MV and SV (see Table 2). These indices show the differences between the normalized most energetic and least energetic months (or seasons).

Another aspect of resource assessment performed for the COWWEF site selection is checking their complementarity. This assessment was conducted by calculating the cross-correlation function (CCF) index between wind and wave power and the event-based approach (see Table 2). Using the CCF index, the correspondence of wind and wave energy resources at time lag t is measured. To avoid interruptions in power generation and to have uniform power output in COWWEF, the CCF at time = 0 (C (0)) and the lag time corresponding to the maximum value of CCF (C max) were used. The lower C (0) and longer lag time were introduced as the best-case scenario (e.g., [38]). Based on the event-based approach, various availability scenarios of wave and wind energy resources were considered by [61] to define four indices (i.e., ECV, VCW, SWV, and UWV). The WCV index was estimated as the frequency of occurrence of the mean annual wind power density above the corresponding lower threshold and the mean annual wave power density below the corresponding lower threshold. The frequency of occurrence of the mean annual wave power density above the corresponding lower threshold and the mean annual wind power density below the corresponding lower threshold was called VCW. The wind and wave power synergy index (SWV) was defined as the frequency of occurrence of the mean annual wave power above the corresponding threshold or the mean annual wind power below the corresponding threshold. The wind and wave power joint non-availability index (UWV) indicates the frequency of occurrence of both mean annual wind and wave power below their corresponding thresholds. The UWV index can be helpful for the exclusion of areas with lower availability of both energy resources, while the others, especially the SWV, reflect the degree of complementarity of wind and wave energy resources.

To check the survivability of energy devices, the 50-year return period of wind speed (WS50), significant wave height (Hs50), current velocity (C50), and water depth (WD) were assessed in the literature [41,59]. Accordingly, the WS50, Hs50, C50, and WD thresholds were used to ensure the WT's structural survivability, while three indices of Hs50, C50, and WD were considered to assess the safety of WECs (see Table 2). In addition, WS50 and Hs50 were considered as EV criteria for site selection of wind and wave energy farms, respectively.

Water Depth (WD) is the main physical parameter impacting the site's suitability. A minimum depth is required based on the structural design of the considered energy-generation device (e.g., draft size) [19]. The (fixed or floating) foundation design [46]; difficulties in cabling layout in deeper water > 100 m [19]; effective design of mooring lines and anchors for floating systems; cost-related issues; and viability of WECs under extreme environmental conditions limit the depth of installation [42]. A relatively limited depth range (e.g., up to 60 m) is adequate for installing fixed-bottom hybrid offshore renewable energy systems [42]. The use of other foundations (e.g., floating systems) has been tested for deeper waters, but there is still a need to refine and develop these designs [50]. As shown in Table 2, the eligible WD range for implementing hybrid offshore wind and wave energy farms is approximately 25–500 m.

Ports provide the infrastructure that enables offshore renewable energy to be installed. The ideal locations are consequently those with the shortest distances to ports due to lower installation, operation, and maintenance costs [40]. Port drafts should be between 10 and 15 m to install offshore renewable energy farms [62]. Constraints for the maximum value of the Distance from Port vary between 50 and 500 m [19,41,42,54]. In some references, only qualitative analysis regarding selecting the feasible areas close to the ports was considered [20,40,47,50]. To check the site accessibility in terms of weather conditions along with DP, two indicators representative of sea state (i.e., WS and Hs) were considered [41].

In some cases, in addition to the visual and noise impact of offshore wind and wave energy farms, looking for an appropriate amount of energy resources leads to moving further from the shore. Nonetheless, the installation and maintenance costs limit the maximum distance from the shore [40]. The evaluated maximum distance range from the

shore in the literature varies between 30 and 444 km (see Table 2). In addition, in some studies, the area close to the shore was selected without mentioning the maximum distance from the shore [20,47,50].

A suitable marine area's score is boosted by its proximity to an electrical grid, as this would eliminate long-distance transmission losses and reduce cabling costs [40]. The maximum distance to the electrical grid is 70 km [40] and 500 km [41]. In some studies, proximity to the high voltage capacity of grid connections was assessed. Based on the available capacity of the local grid, four grid capacities in decreasing preference have thus been defined: 220–400 kV, 220 kV, 150 kV, and 66 kV [42,54]. In addition, the total range of the high capacity of grids in Europe was considered to be between 220 and 500 kV [62].

3.4.2. Economic Criteria

The impact of various feed-in tariffs was assessed by [62] as one of the funding incentives available for offshore wind and wave energy developments. This criterion could be helpful for the study region surrounded by different countries that each offer different prices for a unit of electricity produced.

3.4.3. Socio-Economic Criteria

Meeting the energy demand by installing a hybrid offshore wind–wave renewable energy extraction project is crucial for the project's economic viability and social acceptance. This could be more vital in remote areas such as islands to check the project's potential to cover the region's energy requirements. In [42], the Population Served (PS) criterion was used to serve a municipality's population. It was assumed that if one or more of a municipality's ports were located within a distance less than 100 km from the centroid of an eligible marine area, the prefecture's population could be served. In addition, the average annual electricity consumption was directly considered by [40] as an evaluation criterion.

3.4.4. Environmental Criteria

The lower the environmental impact, the more suitable the location for offshore renewable energy farm developments. In addition to the aforementioned environmentally restricted area for exclusion, other EV criteria have been proposed to reduce the negative environmental impact of COWWEF installations [40,54]. In [54], the environmental performance value (EPV; [67]) was used to explicitly quantify the environmental impacts of COWWEFs. In particular, EPV was calculated by implementing four steps: (i) identifying key environmental components; (ii) assigning a weight of importance to each of the environmental components for two time periods (i.e., existing and future); (iii) assessing the impact significance of the project (i.e., nature of impact including positive or negative, magnitude, permanence, reversibility, and manageability of impact) during different phases of its life cycle (i.e., construction, operation, and decommissioning); and (iv) calculating the EPV. Experts are involved in each step to calculate the EPV for each location. To consider the impact of the project on the abiotic, natural, and anthropogenic environment, 18 components were defined: climate, bioclimate, morphology, aesthetic features, geology, tectonics, soils, natural environment, land uses, built environment, historical and cultural environment, socio-economic environment, infrastructures, atmospheric environment, acoustic environment noise, vibrations, electromagnetic fields, surface waters and groundwater. The EPV criterion was calculated based on the considered value for component weights in Step 2 and defined scaling for assessing the impact significance of the project in Step 3 [54]. The calculated range of EPV is between -164 and 54, which represent extremely negative and extremely positive impacts, respectively. The EPV value thus shows the nature and magnitude of the impact of COWWEF on the aforementioned 19 environmental components, both in the present and future, during three phases of the project life cycle. One of the disadvantages of the EPV criterion is that it cannot be used for prioritizing a large number of locations in a wide study region, as the experts' opinions are required to calculate the mentioned criterion for each location.

To consider the human activities in marine areas and natural habitat vulnerabilities, the Cumulative Impact Index CII was defined by [56], which is expressed as multiplying the Vulnerability Index (VI) by the Cumulative Pressure Index (CPI). These two variables (i.e., VI and CPI) were defined based on the presence or absence and the frequency of occurrences of human activities and vulnerability elements. A low value of the CII represents the site's suitability.

The power-generation equipment would create low-frequency noise, which could negatively affect marine life [68]. Sufficient distance from the marine fauna and flora habitat should be considered to reduce the negative impact on the marine environment and protect nature and wildlife. Hence, distance from the aquaculture farms and nature conservation areas (e.g., natural parks and reserves) was considered by [40] as an EV criterion.

3.4.5. The Frequency of Occurrence of Determinant EV Criteria

All determinant EV criteria (including the criteria related to restrictions given in Figure 3, which have been used for evaluation, and the other relevant criteria mentioned in Table 2), their relevant category, and their frequency of occurrence in the existing studies are listed in Figure 4 to identify which aspect of project deployment has mostly been focused on and which aspect has not been paid attention to or what is the research gap in terms of criteria consideration. As seen from the figure, twenty-seven determinant EV criteria were used for the site selection. Studies were conducted in countries with varied climate and environmental conditions, local policies, and data availability. Hence, different criteria were used in different studies for site selection purposes. In addition, the techno-economic criteria representative of the energy resources, bathymetry, variability and correlation of energy resources, distance to port, shore, and shipping density are the most cited (>5 times) in the literature. In comparison, the evaluation of the environmental impacts of COWWEF has rarely been performed in the literature and is yet to be further investigated.

3.5. Methodologies for Site Selection of Combined Offshore Wind and Wave Energy Farm

Site selection should be multifaceted and include technical, economic, social, and environmental criteria. It is thus a complex decision-making problem that needs systematic analysis of these criteria and the use of appropriate analysis methods. Figure 5 presents a flowchart of the methodologies used for the site selection of COWWEF. Table 3 gives the details of the methods and analysis applied in each study.

3.5.1. Data Collection

As shown in Figure 5, the first step of feasibility analysis is to collect the data (e.g., met-ocean, bathymetry, restricted areas map). After collecting all data, the met-ocean dataset, including wind speed and wave parameters, should be validated to ensure its reliability. Satellite measurements [36] and buoy observations [48] were used to validate the numerical wind and wave model outputs. However, most studies used previously validated numerical data.

3.5.2. Site Selection Method

Two general methods are used for site selection, namely resource-based and MCDM (see Table 3). The former method is focused on the site's energy-generation potential and considers the availability, variability, and correlation of wind and wave energy resources. Less than half of the studies were performed in this way, investigating both variability and correlation of energy resources and the availability of resources. The disadvantage of resource-based assessments is that they do not consider important constraints and socio-environment criteria. In order to consider different conflicting criteria in a decision-making process, MCDM methods are used. These methods have been applied to evaluate different aspects of renewable energy farm developments, including the selection of the best renewable energy technology for a specific area [69], the selection of the best site for the exploitation of specific renewables (the studies reviewed in this paper), and selection of a

suitable energy converter device for a specific area [14]. MCDM site selection methods went beyond considering only energy generation opportunities to cover various other criteria. Important geographical features of the studied region, distance to relevant infrastructure, other sea uses, and various other aspects were considered in the various MCDM studies. Considering that many high-resource-potential COWWEF sites have local constraints and approval challenges, MCDM methods appear more useful for selecting sites that are likely to pass more detailed feasibility assessments and government approval processes.

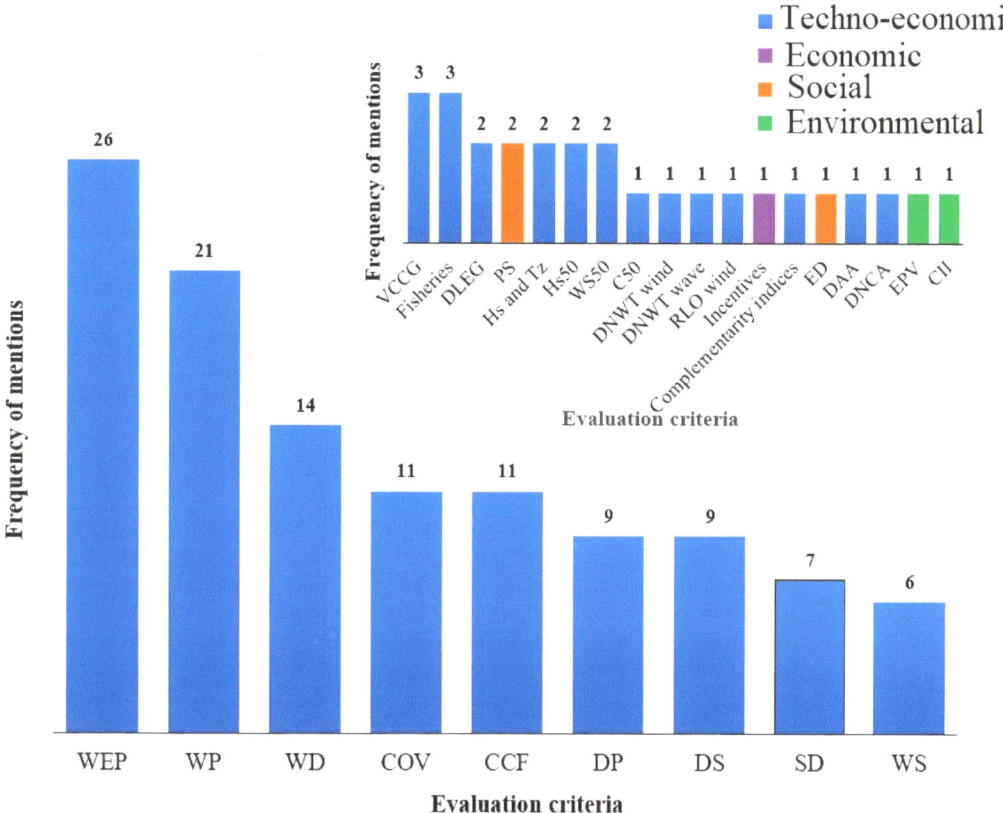

Figure 4. Frequency of occurrence of EV criteria in the literature.

Table 3 and Figure 5 convey that, in some cases, a combination of MCDM methods with GIS or statistical approaches (e.g., Principal Component Analysis (PCA) and Clustering Analysis (CA)) were applied for COWWEF site selection problems. Developers often utilize GIS in many stages of the development process [19]. Using this tool, the overlay maps, which are composed of several layers of information, are analyzed through logical and mathematical operators to determine the ideal location. Many GIS software applications allow the user to customize the application's functions [31]. Combining the MCDM method with GIS could thus facilitate a spatial planning process that would allow the evaluation of numerous site alternatives in an accurate, systematic, and integrated way, thereby reducing subjectivity in decisions [42].

The PCA method enables selecting only a few components to describe the entire dataset with the minimum amount of information loss. Thus, the wind and wave statistics were reduced in dimensionality using PCA in [46,56]. In addition, to analyze the similarities of meteoclimatic data groups, the CA method, including hierarchical (HCA) and

nonhierarchical K-means, was employed in the abovementioned studies. Through these classification methods, the most favorable meteoclimatic conditions in terms of frequency of occurrence [46] and the existing correlation [56] between offshore wind and wave energy resources were identified to be used for site selection.

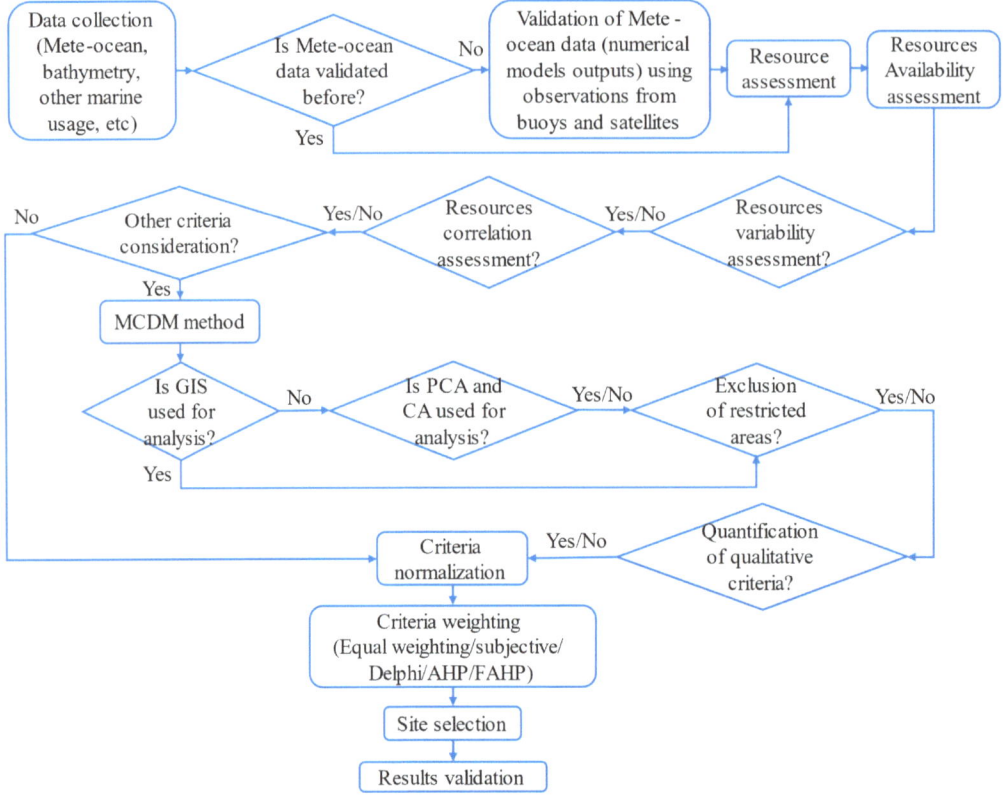

Figure 5. Flow diagram of methodologies used for combined offshore wind–wave energy farm site selection.

3.5.3. Exclusion of Restricted Area and Quantification of the Qualitative Criteria

The existing criteria must either be used as an EX or EV criterion. In most studies, the areas with other marine usage have been excluded (see Section 3.2). Additionally, to conduct site selection analysis, all criteria should be made quantifiable. Qualitative criteria require a verifiable method to transform qualitative assessments into a value that can be combined with other quantitative site selection criteria. Thus, after excluding restricted areas, quantification of qualitative criteria is the next step of COWWEF site selection (see Figure 5). Data quantification was performed in two studies [54,56]. In [56], the area with other marine usages as qualitative criteria was quantified and assigned a score. The impact of offshore renewable energy developments on the environment (i.e., EPV) was quantified by [54] by proposing the EPV criterion. The nature, lingual magnitude, degree of reversibility, permanency, and manageability of the impact were considered to calculate the EPV value.

Table 3. COWWEF site selection analytical methods used in the literature.

Used Method	Data Validation	Variability Correlation	Equal	Subjective Delphi	AHP	FAHP	Normalization	GIS	CA and PCA	Sensitivity	Other	Reference
MCDM	-	✓	✓	-	-	-	-	-	✓	-	-	[46]
	-	✓	-	-	-	-	-	-	-	-	-	[47]
	-	-	-	✓	-	-	-	-	-	-	-	[20]
	✓	✓	-	✓	-	-	✓	✓	-	✓	-	[19]
	✓	✓	-	✓	-	-	✓	-	-	-	-	[48]
	✓	✓	-	✓	-	-	✓	-	-	-	-	[50]
	✓	-	-	-	✓	-	✓	✓	-	✓	-	[42]
	✓	-	-	-	✓	-	✓	✓	-	✓	-	[54]
	✓	-	✓	-	-	✓	✓	-	-	-	✓	[41]
	-	-	✓	-	-	-	-	-	✓	-	-	[56]
	-	✓	-	✓	-	-	✓	-	-	✓	-	[59]
	-	-	-	-	-	-	✓	✓	-	✓	-	[40]
	✓	-	-	✓	-	-	✓	✓	-	-	-	[62]
	✓	✓	-	-	-	✓	-	-	✓	-	-	-	[39]
	-	✓	-	✓	-	-	-	-	-	-	-	[17]
	-	✓	✓	-	-	-	-	-	-	-	-	[60]
	✓	-	-	✓	-	-	-	-	-	-	-	[61]
Resource-based Analysis	✓	✓	-	✓	-	-	-	-	-	-	-	[49]
	✓	✓	-	✓	-	-	-	-	-	-	-	[51]
	✓	✓	-	✓	-	-	-	-	-	-	-	[37]
	✓	✓	✓	-	-	-	-	-	-	-	-	[52]
	-	✓	-	✓	-	-	-	-	-	-	-	[53]

Table 3. *Cont.*

Used Method	Data Validation	Variability	Correlation	Weighting Method				Normalization	GIS	CA and PCA	Results Validation		Reference
				Equal	Subjective Delphi	AHP	FAHP				Sensitivity	Other	
	–	√	√	–	√	–	–	–	–	–	–	–	[55]
	√	√	–	–	√	–	–	–	–	–	–	–	[36]
	–	√	–	√	–	–	–	√	–	–	–	–	[57]
	–	√	√	–	√	–	–	–	–	–	–	√	[58]
	√	√	√	√	–	–	–	√	–	–	–	–	[38]

The signs of √ and – mean that the mentioned method/criterion was used/not used in the corresponding reference.

3.5.4. Normalizing the Criteria Values

In multi-criteria analysis, a decision can be made based on criteria whose units of measure cannot be expressed in the same way. Different criteria magnitudes have different meanings, so in some cases, a high value represents the ideal scenario (i.e., a high value of wind and wave energy); in others, low values (i.e., distance to the shore and port) are better. To conduct the next quantitative operations, it is therefore necessary to convert all variables into directionless and dimensionless operable values. Criteria value normalization or data normalization refers to this process [30]. Table 3 shows that the normalization process was performed in 13 studies involving conflicting criteria and MCDM methods. The reason for not conducting normalization in some studies is that there were no multiple criteria with different units and directions to be normalized, or there was no consideration and comparison of all criteria in one step to select the ideal location. The next step after climate data assessment, exclusion, and quantifying the qualitative criteria is normalizing the criteria (see Figure 5). Normalization allows all criteria to be reduced on a standard scale (0 to 100, 0 to 10, or 0 to 1), with 0 being the worst-case scenario and 1, 10, or 100 being the best-case scenario. The ranking of the criteria values between the worst and the best scenario was conducted based on the number of categories defined for each criterion and the interpolation method.

Extremum processing [30] was also used in some studies [38,39,48,50] as follows:

$$x_{ij}^* = \frac{x_{ij} - m_j}{M_j - m_j} \tag{1}$$

where x_{ij} is the i-th sample of the j-th criterion, x_{ij}^* is the normalized criteria value, and M_j and m_j refer to the maximum and minimum values of the j-th criterion, respectively. The formula below was used in the case of having high priority of the low value of a criterion:

$$x_{ij}^* = \frac{M_j - x_{ij}}{M_j - m_j} \tag{2}$$

The following expression was used by [55,60] to make the used criteria directionless:

$$x_{ij}^* = 1 - x_{ij} \tag{3}$$

Additionally, the linear scaling expressed below was used in both aforementioned studies to make the proposed exploitability index (EI) dimensionless:

$$x_{ij}^* = \frac{x_{ij}}{M_j} \tag{4}$$

In [40], all the criteria values were divided into six categories and were assigned scores of 0 to 5 based on the increasing preference order of criterion values. The percentage of time with favorable energy production conditions and a parametrized function defined for the distance and risk-based criteria were also two methods of normalization used by [41] (see Table 2). The expression of the other method used to normalize the Suitability Index (i.e., the normalized probability range) is [41]

$$PR = \frac{\min(\text{different criteria})}{\max} \times 100 \tag{5}$$

3.5.5. Weighting Method

The next step of the site selection process after normalizing the criteria values is to consider the relative importance of each criterion by weight allocation (see Figure 5). The weights must be rational and accurate. Table 3 displays various weighting methods, including equal weighting, authors' subjectivity, Delphi method, analytical hierarchy process (AHP), and fuzzy AHP (FAHP). Equal weighting is the simplest method, as it

does not require knowledge of the decision-makers' priorities. Nevertheless, it is not the ideal method, as the relative importance of the criteria is ignored. Approximately 26% of existing studies have used this weighting method (see Table 3). In most studies, authors directly considered rank-order weights to the involved criteria based on their own experience and knowledge or on the structural features of hypothetical offshore renewable energy devices (e.g., [19]). These types of weighting are called authors' subjectivity. The subjective weights in some studies are quantitative, while in other studies (i.e., mainly resource-based assessment studies), there are no values for the criteria weights, and the way the authors select the ideal location represents the considered importance of each criterion. Based on the last step of the Delphi classification method applied by [59], which includes a weighting approach to the criteria, experts' opinions were collected to determine the weight coefficients. An average of the assigned weights to each criterion by those who were consulted was thus considered.

The analytic hierarchy process (AHP) initiated by [70] is one of the most popular decision-making techniques in sustainable energy planning and is extensively used for site selection purposes [2,71–74]. Based on Table 3, the AHP approach was applied in two studies [42,54] to calculate the weights of the criteria. This approach facilitates decision-making by providing a mathematical model that assists decision-makers in arriving at the logical choice. By comparing decision criteria pairwise, it uses a quantitative comparison approach that enables accurate weighing of subjective criteria. In AHP, verbal experts' judgments are converted to numbers so that a nine-point scale is used to quantify one option's merit compared to other options on a single scale. In this method, the robustness of the pairwise comparisons is assessed by calculating the consistency ratio (CR; [42]). In the case of inconsistency of pairwise comparisons, namely CR > 0.1, the judgments are therefore modified and repeated, which is the advantage of the AHP method.

To handle the uncertainties in the site selection decision-making process, the fuzzy set theory, introduced by [75], in combination with the MCDM methods, has been employed [76–80]. Presented by [81], FAHP was employed by [40] to obtain relative weights. Utilizing this approach can reduce uncertainties in expert opinions. The fuzzy logic captures how true something is. In this way, the relative importance of items in a pair is judged by decision-makers using linguistic terms. Unlike the AHP method, in which judgments are converted to crisp values, the linguistic experts' opinions are converted into Triangular Fuzzy Numbers (TFNs). Each TFN consists of three terms, indicating the lowest, most probable value, and highest values [82]. A pairwise comparison matrix is therefore created according to the triangular fuzzy conversion scale. The fuzzy comparison matrix is converted into a crisp comparison matrix using the centroid defuzzification method [83]. The consistency of the crisp comparison matrix is then evaluated by calculating the CR as calculated in the AHP method. If inconsistency occurs, a new pairwise comparison judgment is created, and the procession must be continued until consistency is reached. Finally, the fuzzy weights of criteria are obtained using mathematical calculations that are converted into crisp weights using the centroid defuzzification method.

3.5.6. Site Selection

After quantifying and normalizing the criteria and assigning weights to each criterion, the final steps of the analysis are the selection of appropriate locations and ranking of the feasible areas. The weighted overlay approach was used to rank the suitability of the existing areas (e.g., [40,42,54]). Using the Weighted Linear Combination (WLC), the weighted criteria values of each location are summed to obtain a value representing the suitability of each site. However, based on the subjective weighting of most studies, the ranking was conducted step-by-step, from considering the most important criterion for the selection of a suitable location to the least important one (e.g., [20,47,58,61]). First, the preferable area was chosen based on the most important criterion, and the other criteria values were then checked in that location to select the final location.

3.5.7. Results Validation

Due to the inherent uncertainty that characterizes most decision-making processes, the result validation is another crucial step when conflicting criteria exist and MCDM methods are used (see Figure 5 and Table 3). This stage is performed to ensure the reliability of the results and final decision. Result validation consists of sensitivity analysis or other types of validation. By changing the criteria limits [19] or their assigned weights (e.g., [40,42,54,59]), the sensitivity analysis reveals a new result that should be compared with the previous one. In another validation method, the results are compared with those of other studies [41,58]. The effect of varying limitations of decision criteria (e.g., distance to the port and environmental exclusions) was investigated by [19]. This sensitivity analysis was performed due to the limited data available and the lack of a clear definition of the specific limits for the criteria (see Table 3). In addition, a particular sensitivity analysis method, namely the Monte Carlo approach, was used to evaluate the impact of the variability of the weight coefficients on the final decision [59]. With this method, the new weights were generated considering random numbers, previous weights assigned to each criterion, and their standard deviation. After random variation of weights, they were renormalized to ensure their summation was equal to 1, and the new weights were calculated. Finally, the defined classification index for identifying the ideal location was calculated according to the new weights, and the results were compared with the previous ones.

3.6. Key Challenges

- The lack of met-ocean data with high resolution in most offshore areas around the world is one of the challenges researchers face during the site selection process. Although the global wind and wave models are available with coarse resolution, the detailed feasibility analysis of combined power plant installations at a local level requires data with fine resolution produced by running the numerical models. On the other hand, downscaling the data is always associated with uncertainty and errors, especially with the lack of in situ measurements. Therefore, the output of these models is not perfect.
- Considering the uncertainty involved in the site selection process, which originates from the limited understanding of the problem, inconsistency in expert opinions and the stochastic feature of sea state and climate condition is another challenge that researchers should address. It should be mentioned that the uncertainty of experts' options for weighting the criteria was reduced using the FAHP method [40] to select the optimal sites for COWWEF developments.
- Environmental restrictions have been widely considered in the literature to exclude vulnerable areas from potential sites for COWWEF development. Nevertheless, there are still some environmental components that can physically or biologically be affected by energy devices. Therefore, assessing the possible environmental impacts of marine renewable energy farms remains challenging. The proposed EPV criterion by [54,67] can only be used to prioritize a limited number of locations for COWWEF developments, as the experts' knowledge is required to calculate this criterion for each location.
- Regarding the long lifespan of energy devices, climate change can impact the results of site selection analysis. For example, the change in the sea state and climate condition directly affects the potential energy resources in a region which may lead to different optimal locations for device installations. In addition, the water depth is affected by sea level rise and coastal erosions are caused by climate change. Although the impact of climate change on resources potential for site selection of COWWEF has rarely been evaluated [60], incorporating the variation of input criteria involved in the site selection procedure as a result of climate change is a challenging issue.

4. Conclusions and Recommendations for Future Work

This study provided a systematic literature review with a critical and comprehensive analysis of publications to identify the restrictions and the relevant criteria for determining the best location for COWWEF development. Furthermore, this review contributed to understanding the applied methods and datasets required for evaluating a project's production capacity. Based on the obtained results, the main findings of this paper are summarized below:

- Regarding study regions, mainly European offshore areas were (85%) evaluated to select the appropriate location for COWWEF.
- Most of the literature (93%) relies on numerical models to provide long-term datasets with high spatial and temporal resolution due to the lack of measurements from buoys and satellites.
- The restrictions, which were considered based on local laws, are related to marine and environmental usage. A total of 12 restricted areas were identified which featured shipping lanes, military exercise, and marine protected areas, and the areas close to the shore were the most frequently listed.
- Twenty-seven EV criteria were identified from various technical, economic, social, and environmental perspectives. Among EV criteria, those representing wind and wave energy resource potential, bathymetry, variability, and correlation of wind and wave energy resources, as well as distance to infrastructures such as ports, were the most frequently considered.
- Two approaches, namely MCDM (14 publications) and resource-based (13 studies), were applied to select the optimal sites for locating COWWEF. The GIS and statistical approaches, including PCA and CA, were also used in the literature in combination with MCDM methods.

Based on the performed review, some research gaps which highlight the direction for future research were identified as follows (see also Figure 6 for a summary of identified research gaps and recommendations for future works):

- The literature lacks a consensus on using all the relevant criteria for the site selection process. A comprehensive framework for this purpose is yet to be established.
- While most studies rely on other aspects, especially the techno-economic aspects of COWWEF site selection, more emphasis should be placed on the environmental impacts of the project development. This includes identifying and incorporating biological/physical impacts on the local environmental components in different phases of the project (i.e., construction, operation and maintenance, and decommissioning) into a decision-making analysis. The mentioned impacts can be considered based on experts' knowledge about the studied region's specific ecosystem diversity.
- The seabed's physical characteristics significantly influence the project's cost in terms of the constructability of the structural foundation or deployment of the mooring cables. Sandy seabed and mild seafloor slopes are preferred as a rock-covered or steep slope can significantly increase costs. Future research should focus on improving input parameters considering geospatial economics, including seabed type and slope.
- Market analysis has rarely been considered in the literature. Utility feed-in tariffs can be an efficient input parameter for the decision-making process, especially when the studied offshore area is surrounded by several countries or states offering different prices for the electricity generated from certain sources.
- The site selection process is associated with a high level of uncertainty. The FAHP used in the literature only reduces the uncertainty of experts' opinions for weighting the criteria [40]. Bayesian Network (BN) [84], which considers probabilities, is a suitable option for decision-making under uncertainty. It is aimed at solving problems with uncertainty due to inconsistency in the knowledge of experts, limited understanding of the problem, or stochastic phenomena [85]. The feature of scenario analysis using BN makes it useful to formulate probabilistic changes in the future [86,87].

- Investigating the future exploitability of wind and wave energy resources is crucial, considering the long lifespan of energy-generation devices. This issue has rarely been addressed in the literature. Investigating the impact of climate change on the selection of optimal locations for COWWEF developments is recommended for future studies.

Research gaps	Recommendations
Lack of consensus about using all relevant criteria in site selection process.	A Comprehensive study considering all technical, economic, environmental, and social aspects of COWWEF site selection should be conducted.
Limited consideration of the impact of COWWEF development on the local environmental components	The possible negative impacts of COWWEF on local ecosystem diversity should be considered in future studies based on experts' knowledge.
Lack of consideration of seabed geology and slope for locating COWWEF.	Future research should focus on improving input parameters that consider geospatial economics, including seabed type and slope.
Limited market analysis for positioning COWWEF	Utility feed-in tariff offered by countries/states involved in studied offshore region should be considered as an contributing factor in site selection process.
Lack of consideration of the uncertainties associated with COWWEF site selection process.	Decision support techniques that handle the inherent uncertainty associated with input parameters in COWWEF site selection (e.g. BN) should be applied.
Scarcity of investigations on the future exploitability of combined wind and wave energy resources.	COWWEF site selection approaches should account for climate change.

Figure 6. A summary of research gaps and recommendations for future studies.

Researchers and decision-makers considering offshore renewable energy solutions will benefit from this comprehensive review study since it catalogs and synthesizes all information about COWWEF site selection and highlights some recommendations for future studies. This work aids researchers and practitioners in selecting the best COWWEF sites, ultimately helping society to transition to a renewable energy future.

Author Contributions: Conceptualization, S.H., A.E.-S. and R.A.S.; methodology, S.H., A.E.-S. and R.A.S.; formal analysis, S.H., A.E.-S. and R.A.S.; investigation, S.H., A.E.-S. and R.A.S.; resources, S.H., A.E.-S. and R.A.S.; writing—original draft preparation, S.H.; writing—review and editing, S.H., A.E.-S. and R.A.S.; visualization, S.H., A.E.-S. and R.A.S.; supervision, A.E.-S. and R.A.S.; funding acquisition, S.H. All authors have read and agreed to the published version of the manuscript.

Funding: This research was supported by Griffith University scholarships (GUPRS and GUIPRS) awarded to the first author to pursue her Ph.D.

Data Availability Statement: Not applicable.

Acknowledgments: The authors are grateful for the support of Griffith University and CRI in providing facilities and resources and awarding GUPRS and GUIPRS scholarships to the first author to pursue her Ph.D.

Conflicts of Interest: The authors declare no conflict of interest.

Nomenclature

AHP	Analytic Hierarchy Process
BN	Bayesian Network
CA	Clustering Analysis
CAWCR	Center for Australian Weather and Climate Research
CFSR	Climate Forecast System Reanalysis
CORDEX	Coordinated Regional Downscaling Experiment
COWWEF	Combined Offshore Wind and Wave Energy Farm
ECMWF	European Center for Medium-Range Weather Forecasts
EV	Evaluation
EX	Exclusion
FAHP	Fuzzy Analytic Hierarchy Process
GCM	Global Climate Model
GIS	Geographic Information System
MCDM	Multi-Criteria Decision-Making
NASA	National Aeronautics and Space Administration
NCEP	National Center for Environmental Prediction
PCA	Principal Component Analysis
RCM	Regional Climate Model
RCP	Representative Concentration Pathway
REMO	Regional Climate Model
SLR	Systematic Literature Review
SWAN	Simulating Waves Near-shore
TFN	Triangular Fuzzy Numbers
WAM	Wave Model
WAsP	Wind Atlas Analysis and Application Program
WRF	Weather Research and Forecasting
WT	Wind Turbine
WEC	Wave Energy Converter

Appendix A

Table A1. Details of criteria, including definitions, formulae, and acceptable range.

Category	Sub-Category	Criteria, Parameters, and Definitions	Acceptable Range	No.
Techno-Economic	Wind energy resource richness	Wind Power, WP (W/m^2) = $\frac{1}{2}\rho V^3$ V: Wind Speed, ρ: Air density	WP $\geq$ 280	1
			WP $\geq$ 50	1
			-	18
		Wind Speed, WS (m/s)	$6 \leq$ WS ≤ 8	2
			WS (Annual average wind speed)	1
			WS $\geq$ 6–7	2
			-	1
		SI $_{\text{Wind R}}$ = min (Ap, $\frac{t_{Hs}}{\bar{t}}$) Ap = $\begin{cases} \frac{t_{WP}}{\bar{t}} & for \ \frac{t_{WP}}{\bar{t}} < 0.7 \\ 1 & for \ \frac{t_{WP}}{\bar{t}} \geq 0.7 \end{cases}$ t_{Hs} and t_{WP} are the time in which Hs (m) and WP (W/m^2) are respectively in the acceptable range $\bar{t}$= The total time of the data series (s)	$0 \leq$ SI $_{\text{Wind R}} \leq 1$ WP $\geq$ 400 Hs $\leq$ 5	1

Table A1. Cont.

Category	Sub-Category	Criteria, Parameters, and Definitions	Acceptable Range	No.
		$DWNT_{wind}$ (%) Total time with useful wind speed ($4 \leq WS \leq 25$ (m/s)) in which wind turbine is producing electricity	$DWNT_{WI} \geq 10$ $4 \leq WS \leq 25$	1
		RLO_{wind} (%) Rich Level Occurrence = Frequency of wind power higher than 200 W/m² (WP > 200 W/m²)	$RLO_{WI} \geq 10$ WP > 200	1
	Wave energy resource richness	Wave Energy Power, WEP (kW/m) $WEP = \frac{\rho_w g^2 H_s^2 T_e}{64\pi}$ ρ_w is the seawater density, g is gravitational acceleration, H_s is significant wave height, T_e is the mean wave period	$5 \leq WEP \leq 10$	2
			$WEP \geq 5$	2
			$WEP \geq 2$	1
			$WEP \geq 10$	1
			-	18
			$WEP \geq 20\text{-}30$	1
		Significant wave height, Hs (m), and wave period, Tz (s)	-	2
		$SI_{Wave\,R} = \frac{\left(\left(\frac{t_{WEP}}{\bar{t}} * 2\right) + \frac{t_{Hs}}{\bar{t}} + \frac{t_{Tp}}{\bar{t}}\right)}{4}$ t_{WEP}, t_{Hs}, t_{Tp} are the time in which the WEP, Hs, and Tp are respectively in the acceptable range $\bar{t}$ = The total time of the data series	$0 \leq SI_{Wave\,R} \leq 1$ $WEP \geq 15$ $1 \leq H_s \leq 6$ $5 \leq T_p \leq 14$	1
		$DWNT_{Wave\,(\%)}$ = Total time with useful significant wave height ($1 \leq Hs \leq 8$ m) in which wave energy converter is producing electricity	$DWNT_{WA} \geq 10$	1
	Combined offshore wind-wave farm richness	Mean Capacity Factor of combined energy farm, CF_{comb}	-	1
		Downtime or non-production time of combined energy farm, DT	-	1
	Resource variability	Coefficient of Variation, COV Or Total Harmonic Distortion, THD	-	10
			$COV \leq 1.9$	1
		Monthly Variation, $MV = \frac{P_{Mmax} - P_{Mmin}}{P_{year}}$	-	4
			$MV \leq 2.5$	1
		Seasonal Variation, $SV = \frac{P_{Smax} - P_{Smin}}{P_{year}}$	-	4
		Skewness, S	-	4
		Kurtosis, K	-	3
		Standard Deviation, SD	-	4
	Resource complementarity	Cross-Correlation Factor, CCF	-	11
		Wind-to-wave-power Complementarity Index (WCV), Wave-to-wind-power Complementarity Index (VCW), Synergy Index (SWV), Joint Non-availability Index (UWV)	-	1

Table A1. Cont.

Category	Sub-Category	Criteria, Parameters, and Definitions	Acceptable Range	No.
	Devices' survivability	$SI_{Wind\,S} = \min(f(H_{s50}), f(WS_{50}), f(C_{50}))$ $f(x) = \begin{cases} \frac{-0.8}{thld}x + 1 & for\ x \leq thld \\ \frac{0.2(x-max)}{thld-max} & for\ x > thld \end{cases}$ First, the acceptable range of WD was considered. Then, the $SI_{Wind\,S}$ was calculated based on the thresholds given in the acceptable range column for each used parameter Hs_{50} (m), WS_{50} (m/s), and C_{50} (m/s) are 50-year return periods of significant wave height, wind speed, and current velocity, respectively	$0 \leq SI_{Wind\,S} \leq 1$ $WS_{50} \leq 40$ $Hs_{50} \leq 15$ $C_{50} \leq 2$ $WD \leq 500$	1
		WS_{50}	$WS_{50} \leq 27$	1
		$SI_{Wave\,S} = \min(f(H_{s50}), f(C_{50}))$ $f(x) = \begin{cases} \frac{-0.8}{thld}x + 1 & for\ x \leq thld \\ \frac{0.2(x-max)}{thld-max} & for\ x > thld \end{cases}$ First, the acceptable range of WD was considered. Then, the $SI_{Wave\,S}$ was calculated based on the thresholds given in the acceptable range column for each used parameter	$0 \leq SI_{Wave\,S} \leq 1$ $Hs_{50} \leq 15$ $C_{50} \leq 2$ $WD \leq 500$	1
		H_{s50}	$H_{s50} \leq 21$	1
	Foundation/anchoring design of devices	Water Depth, WD (m)	$WD \leq 500$	4
			$WD \leq 300$	1
			$70\text{-}150 \leq WD \leq 250$	1
			$WD \leq 35\text{--}50$	1
			$25 \leq WD \leq 100\ (50)$	1
			$WD \leq 100$	1
			$WD \leq 50$	2
			$35 \leq WD \leq 75$	1
			$50 \leq WD \leq 350$	1
			-	1
	Logistics (Feasibility of installation, operation, and maintenance)	$SI_{Log} = \min(\frac{t_{WS}}{\bar{t}}, \frac{t_{Hs}}{\bar{t}}, f(DP))$ $f(x) = \begin{cases} \frac{-0.8}{thld}x + 1 & for\ x \leq thld \\ \frac{0.2(x-max)}{thld-max} & for\ x > thld \end{cases}$ The threshold of DP was given in the column representing an acceptable range of parameters t_{WS}, t_{Hs} the time in which the WS and Hs are respectively in the acceptable range $\bar{t}$ = The total time of the data series	$0 \leq SI_{Log} \leq 1$ $WS_{50} \leq 10$ $Hs_{50} \leq 2$ $DP \leq 250$	1

Table A1. Cont.

Category	Sub-Category	Criteria, Parameters, and Definitions	Acceptable Range	No.
		Distance from Port, DP (km)	DP $\leq$ 250	1
			DP $_{(O\&M)}$ $\leq$ 50–100–200 DP $_{(Construction)}$ $\leq$ 200–500 DP$_{(O\&M)}$ $\leq$100 & DP$_{(Construction)}$ $\leq$ 500	1
			50 $\leq$ DP $\leq$ 100	2
			-	4
			DP (Deep water) $\leq$ 500 DP (Shallow water) $\leq$ 130	1
	Transmission cost and energy dissipation	Distance from Shore, DS (km)	-	3
			DS = 100	1
			DS $\leq$ 30	2
			DS $\leq$ 444	1
			DS $\leq$ 50–100–150	1
			DS $\leq$ 200	1
		Distance to the Local Electrical Grid, DLEG (km)	DLEG $\leq$ 500	1
			DLEG $\leq$ 70	1
		Voltage Capacity of Closest available Grid, VCCG (kv)	66–400	2
			220–500	1
Economic	Prioritizing different countries based on the feed-in tariff	Incentives: Feed-in tariffs of different countries located around the study region	-	1
Socio-Economic	Supplying energy demand	Population Served, PS	DP $\leq$ 100	2
		Electricity Demand, ED: A candidate area's electricity demand was estimated based on the local province's average annual electricity consumption	-	1
Environmental	Impact of offshore renewable energy farms on the environment	Environmental Performance Value, EPV	$-162 \leq$ EPV ≤ 54 -162: Extremely negative 54: Extremely positive	1
	Impact of considering human activities and environmental vulnerability in the site selection process	Cumulative Impact Index, CII = Multiplying the Cumulative Pressure Index (CPI) by the Vulnerability Index (VI)	$0 \leq 0.04$, Low 0.05–0.33, Moderate 0.34–0.61, High >0.62, Very high	1
	Impact of noise on the growth of marine animals due to low frequency of sound waves	Distance from Aquaculture Area, DAA (km)	≥ 1	1
	Impact of hitting birds by turbine blades	Distance from Nature Conservation Areas, DNCA (km): include natural parks, natural reserves, flora and fauna habitats that protect nature and wildlife	≥ 1	1

References

1. UNFCCC. The Glasgow Climate Pact—Key Outcomes from COP26. unfccc.int. Available online: https://unfccc.int/process-and-meetings/the-paris-agreement/the-glasgow-climate-pact-key-outcomes-from-cop26 (accessed on 1 July 2022).
2. Emeksiz, C.; Demirci, B. The Determination of Offshore Wind Energy Potential of Turkey by Using Novelty Hybrid Site Selection Method. *Sustain. Energy Technol. Assess.* **2019**, *36*, 100562. [CrossRef]
3. Bp. *Bp's Statistical Review of World Energy 2021*, 70th ed.; Whitehouse Associates: London, UK, 2021; p. 5.
4. Mohammadi, S.; Maleki, A.; Ehsani, R.; Shakouri, O. Investigation of Wind Energy Potential in Zanjan Province, Iran. *Renew. Energy Res. Appl.* **2022**, *3*, 61–70. [CrossRef]
5. Irena IREA. Statistics Time Series. 2022. Available online: https://www.irena.org/Statistics/View-Data-by-Topic/Capacity-and-Generation/Statistics-Time-Series (accessed on 1 June 2022).
6. Golestani, N.; Arzaghi, E.; Abbassi, R.; Garaniya, V.; Abdussamie, N.; Yang, M. The Game of Guwarra: A Game Theory-Based Decision-Making Framework for Site Selection of Offshore Wind Farms in Australia. *J. Clean. Prod.* **2021**, *326*, 129358. [CrossRef]
7. Sun, X.; Huang, D.; Wu, G. The Current State of Offshore Wind Energy Technology Development. *Energy* **2012**, *41*, 298–312. [CrossRef]
8. Li, Y.; Huang, X.; Tee, K.F.; Li, Q.; Wu, X.P. Comparative Study of Onshore and Offshore Wind Characteristics and Wind Energy Potentials: A Case Study for Southeast Coastal Region of China. *Sustain. Energy Technol. Assess.* **2020**, *39*, 100711. [CrossRef]
9. Deveci, M.; Özcan, E.; John, R.; Covrig, C.F.; Pamucar, D. A Study on Offshore Wind Farm Siting Criteria Using a Novel Interval-Valued Fuzzy-Rough Based Delphi Method. *J. Environ. Manag.* **2020**, *270*, 110916. [CrossRef] [PubMed]
10. Breton, S.P.; Moe, G. Status, Plans and Technologies for Offshore Wind Turbines in Europe and North America. *Renew. Energy* **2009**, *34*, 646–654. [CrossRef]
11. Khan, M.; Khalid, A.; Lughmani, W.A.; Khan, M.M. A Use Case of Exclusive Economic Zone of Pakistan for Wave Power Potential Estimation. *Ocean Eng.* **2021**, *237*, 109664. [CrossRef]
12. Dunnett, D.; Wallace, J.S. Electricity Generation from Wave Power in Canada. *Renew. Energy* **2009**, *34*, 179–195. [CrossRef]
13. Rusu, E. Wave Energy Assessments in the Black Sea. *J. Mar. Sci. Technol.* **2009**, *14*, 359–372. [CrossRef]
14. Kamranzad, B.; Hadadpour, S. A Multi-Criteria Approach for Selection of Wave Energy Converter/Location. *Energy* **2020**, *204*, 117924. [CrossRef]
15. Corsatea, T.D.; Magagna, D. *Overview of European Innovation Activities in Marine Energy Technology JRC Science and Policy Reports*; Publications Office of the European Union: Luxembourg, 2013.
16. Leijon, M.; Danielsson, O.; Eriksson, M.; Thorburn, K.; Bernhoff, H.; Isberg, J.; Sundberg, J.; Ivanova, I.; Sjöstedt, E.; Ågren, O.; et al. An Electrical Approach to Wave Energy Conversion. *Renew. Energy* **2006**, *31*, 1309–1319. [CrossRef]
17. Fusco, F.; Nolan, G.; Ringwood, J.V. Variability Reduction through Optimal Combination of Wind/Wave Resources—An Irish Case Study. *Energy* **2010**, *35*, 314–325. [CrossRef]
18. Rashidi, M.M.; Mahariq, I.; Murshid, N.; Wongwises, S.; Mahian, O.; Alhuyi Nazari, M. Applying Wind Energy as a Clean Source for Reverse Osmosis Desalination: A Comprehensive Review. *Alex. Eng. J.* **2022**, *61*, 12977–12989. [CrossRef]
19. Cradden, L.; Kalogeri, C.; Barrios, I.M.; Galanis, G.; Ingram, D.; Kallos, G. Multi-Criteria Site Selection for Offshore Renewable Energy Platforms. *Renew. Energy* **2016**, *87*, 791–806. [CrossRef]
20. Veigas, M.; Carballo, R.; Iglesias, G. Wave and Offshore Wind Energy on an Island. *Energy Sustain. Dev.* **2014**, *22*, 57–65. [CrossRef]
21. Erdinc, O.; Uzunoglu, M. Optimum Design of Hybrid Renewable Energy Systems: Overview of Different Approaches. *Renew. Sustain. Energy Rev.* **2012**, *16*, 1412–1425. [CrossRef]
22. Jones, D.F.; Wall, G. An Extended Goal Programming Model for Site Selection in the Offshore Wind Farm Sector. *Ann. Oper. Res.* **2016**, *245*, 121–135. [CrossRef]
23. Nobre, A.; Pacheco, M.; Jorge, R.; Lopes, M.F.P.; Gato, L.M.C. Geospatial Multi-Criteria Analysis for Wave Energy Conversion System Deployment. *Renew. Energy* **2009**, *34*, 97–111. [CrossRef]
24. Tsoutsos, T.; Drandaki, M.; Frantzeskaki, N.; Iosifidis, E.; Kiosses, I. Sustainable Energy Planning by Using Multi-Criteria Analysis Application in the Island of Crete. *Energy Policy* **2009**, *37*, 1587–1600. [CrossRef]
25. Gkeka-Serpetsidaki, P.; Tsoutsos, T. A Methodological Framework for Optimal Siting of Offshore Wind Farms: A Case Study on the Island of Crete. *Energy* **2022**, *239*, 122296. [CrossRef]
26. Keivanpour, S.; Ramudhin, A.; Ait Kadi, D. The Sustainable Worldwide Offshore Wind Energy Potential: A Systematic Review. *J. Renew. Sustain. Energy* **2017**, *9*, 065902. [CrossRef]
27. Gil-García, I.C.; García-Cascales, M.S.; Fernández-Guillamón, A.; Molina-García, A. Categorization and Analysis of Relevant Factors for Optimal Locations in Onshore and Offshorewind Power Plants: A Taxonomic Review. *J. Mar. Sci. Eng.* **2019**, *7*, 391. [CrossRef]
28. Peters, J.L.; Remmers, T.; Wheeler, A.J.; Murphy, J.; Cummins, V. A Systematic Review and Meta-Analysis of GIS Use to Reveal Trends in Offshore Wind Energy Research and Offer Insights on Best Practices. *Renew. Sustain. Energy Rev.* **2020**, *128*, 109916. [CrossRef]
29. Spyridonidou, S.; Vagiona, D.G. Systematic review of site-selection processes in onshore and offshore wind energy research. *Energies* **2020**, *13*, 5906. [CrossRef]

30. Shao, M.; Han, Z.; Sun, J.; Xiao, C.; Zhang, S.; Zhao, Y. A Review of Multi-Criteria Decision Making Applications for Renewable Energy Site Selection. *Renew. Energy* **2020**, *157*, 377–403. [CrossRef]
31. Rediske, G.; Burin, H.P.; Rigo, P.D.; Rosa, C.B.; Michels, L.; Siluk, J.C.M. Wind Power Plant Site Selection: A Systematic Review. *Renew. Sustain. Energy Rev.* **2021**, *148*, 111293. [CrossRef]
32. Clemente, D.; Rosa-Santos, P.; Taveira-Pinto, F. On the Potential Synergies and Applications of Wave Energy Converters: A Review. *Renew. Sustain. Energy Rev.* **2021**, *135*, 110162. [CrossRef]
33. Pérez-Collazo, C.; Greaves, D.; Iglesias, G. A Review of Combined Wave and Offshore Wind Energy. *Renew. Sustain. Energy Rev.* **2015**, *42*, 141–153. [CrossRef]
34. Wassie, Y.T.; Adaramola, M.S. Potential Environmental Impacts of Small-Scale Renewable Energy Technologies in East Africa: A Systematic Review of the Evidence. *Renew. Sustain. Energy Rev.* **2019**, *111*, 377–391. [CrossRef]
35. Babatunde, K.A.; Begum, R.A.; Said, F.F. Application of Computable General Equilibrium (CGE) to Climate Change Mitigation Policy: A Systematic Review. *Renew. Sustain. Energy Rev.* **2017**, *78*, 61–71. [CrossRef]
36. Rusu, E.; Onea, F. Joint Evaluation of the Wave and Offshore Wind Energy Resources in the Developing Countries. *Energies* **2017**, *10*, 1866. [CrossRef]
37. Rusu, E.; Onea, F. A Parallel Evaluation of the Wind and Wave Energy Resources along the Latin American and European Coastal Environments. *Renew. Energy* **2019**, *143*, 1594–1607. [CrossRef]
38. Gao, Q.; Saeed Khan, S.; Sergiienko, N.; Ertugrul, N.; Hemer, M.; Negnevitsky, M.; Ding, B. Assessment of Wind and Wave Power Characteristic and Potential for Hybrid Exploration in Australia. *Renew. Sustain. Energy Rev.* **2022**, *168*, 112747. [CrossRef]
39. Patel, R.P.; Nagababu, G.; Kachhwaha, S.S.; Kumar, S.V.V.A.; Seemanth, M. Combined Wind and Wave Resource Assessment and Energy Extraction along the Indian Coast. *Renew. Energy* **2022**, *195*, 931–945. [CrossRef]
40. Zhou, X.; Huang, Z.; Wang, H.; Yin, G.; Bao, Y.; Dong, Q.; Liu, Y. Site Selection for Hybrid Offshore Wind and Wave Power Plants Using a Four-Stage Framework: A Case Study in Hainan, China. *Ocean Coast. Manag.* **2022**, *218*, 106035. [CrossRef]
41. Weiss, C.V.C.; Guanche, R.; Ondiviela, B.; Castellanos, O.F.; Juanes, J. Marine Renewable Energy Potential: A Global Perspective for Offshore Wind and Wave Exploitation. *Energy Convers. Manag.* **2018**, *177*, 43–54. [CrossRef]
42. Vasileiou, M.; Loukogeorgaki, E.; Vagiona, D.G. GIS-Based Multi-Criteria Decision Analysis for Site Selection of Hybrid Offshore Wind and Wave Energy Systems in Greece. *Renew. Sustain. Energy Rev.* **2017**, *73*, 745–757. [CrossRef]
43. Kaldellis, J.K.; Apostolou, D.; Kapsali, M.; Kondili, E. Environmental and Social Footprint of Offshore Wind Energy. Comparison with Onshore Counterpart. *Renew. Energy* **2016**, *92*, 543–556. [CrossRef]
44. Serri, L.; Sempreviva, A.; Pontes, T.; Murphy, J.; Lynch, K.; Airoldi, D.; Hussey, J.; Rudolph, C.; Karagali, I. *Resource Data and GIS Tool for Offshore Renewable Energy Projects in Europe*; Results FP7 ORECCA Proj. Work Packag. 2; Hydraulics & Maritime Research Centre: Cork, Ireland, 2012.
45. Lichtenwalner, S. Ocean Data Labs. Available online: https://datalab.marine.rutgers.edu/2013/07/satellites-vs-buoys/ (accessed on 10 May 2022).
46. Azzellino, A.; Ferrante, V.; Kofoed, J.P.; Lanfredi, C.; Vicinanza, D. Optimal Siting of Offshore Wind-Power Combined with Wave Energy through a Marine Spatial Planning Approach. *Int. J. Mar. Energy* **2013**, *3–4*, e11–e25. [CrossRef]
47. Veigas, M.; Iglesias, G. Wave and Offshore Wind Potential for the Island of Tenerife. *Energy Convers. Manag.* **2013**, *76*, 738–745. [CrossRef]
48. Astariz, S.; Iglesias, G. Selecting Optimum Locations for Co-Located Wave and Wind Energy Farms. Part I: The Co-Location Feasibility Index. *Energy Convers. Manag.* **2016**, *122*, 589–598. [CrossRef]
49. Rusu, E. The Synergy between Wind and Wave Power along the Coasts of the Black Sea. *Marit. Transp. Harvest. Sea Resour.* **2018**, *2*, 1211–1217.
50. Astariz, S.; Iglesias, G. The Collocation Feasibility Index—A Method for Selecting Sites for Co-Located Wave and Wind Farms. *Renew. Energy* **2017**, *103*, 811–824. [CrossRef]
51. Onea, F.; Ciortan, S.; Rusu, E. Assessment of the Potential for Developing Combined Wind-Wave Projects in the European Nearshore. *Energy Environ.* **2017**, *28*, 580–597. [CrossRef]
52. Kalogeri, C.; Galanis, G.; Spyrou, C.; Diamantis, D.; Baladima, F.; Koukoula, M.; Kallos, G. Assessing the European Offshore Wind and Wave Energy Resource for Combined Exploitation. *Renew. Energy* **2017**, *101*, 244–264. [CrossRef]
53. Rusu, L.; Ganea, D.; Mereuta, E. A Joint Evaluation of Wave and Wind Energy Resources in the Black Sea Based on 20-Year Hindcast Information. *Energy Explor. Exploit.* **2018**, *36*, 335–351. [CrossRef]
54. Loukogeorgaki, E.; Vagiona, D.G.; Vasileiou, M. Site Selection of Hybrid Offshore Wind and Wave Energy Systems in Greece Incorporating Environmental Impact Assessment. *Energies* **2018**, *11*, 2095. [CrossRef]
55. Rusu, L. The Wave and Wind Power Potential in the Western Black Sea. *Renew. Energy* **2019**, *139*, 1146–1158. [CrossRef]
56. Azzellino, A.; Lanfredi, C.; Riefolo, L.; Contestabile, P.; Vicinanza, D. Combined Exploitation of Offshore Wind and Wave Energy in the Italian Seas: A Spatial Planning Approach. *Front. Energy Res.* **2019**, *7*, 42. [CrossRef]
57. Ferrari, F.; Besio, G.; Cassola, F.; Mazzino, A. Optimized Wind and Wave Energy Resource Assessment and Offshore Exploitability in the Mediterranean Sea. *Energy* **2020**, *190*, 116447. [CrossRef]

58. Gaughan, E.; Fitzgerald, B. An Assessment of the Potential for Co-Located Offshore Wind and Wave Farms in Ireland. *Energy* **2020**, *200*, 117526. [CrossRef]
59. Ribeiro, A.; Costoya, X.; de Castro, M.; Carvalho, D.; Dias, J.M.; Rocha, A.; Gomez-Gesteira, M. Assessment of Hybrid Wind-Wave Energy Resource for the NW Coast of Iberian Peninsula in a Climate Change Context. *Appl. Sci.* **2020**, *10*, 7359. [CrossRef]
60. Lira-Loarca, A.; Ferrari, F.; Mazzino, A.; Besio, G. Future Wind and Wave Energy Resources and Exploitability in the Mediterranean Sea by 2100. *Appl. Energy* **2021**, *302*, 117492. [CrossRef]
61. Kardakaris, K.; Boufidi, I.; Soukissian, T. Offshore Wind and Wave Energy Complementarity in the Greek Seas Based on ERA5 Data. *Atmosphere* **2021**, *12*, 1360. [CrossRef]
62. Murphy, J.; Lynch, K.; Serri, L.; Airdoldi, D.; Lopes, M. Site Selection Analysis for Offshore Combined Resource Projects in Europe. *Orecca* **2011**, *123*, 1–117.
63. Van der Wal, J.T.; Quirijns, F.J.; Leopold, M.F.; Slijkerman, D.M.E.; Jongbloed, R.H. *Identification and Analysis of Interations between Sea Use Functions*; Wageningen University & Research Publications: Wageningen, The Netherlands, 2009.
64. Costoya, X.; DeCastro, M.; Santos, F.; Sousa, M.C.; Gómez-Gesteira, M. Projections of Wind Energy Resources in the Caribbean for the 21st Century. *Energy* **2019**, *178*, 356–367. [CrossRef]
65. Costoya, X.; deCastro, M.; Carvalho, D.; Gómez-Gesteira, M. On the Suitability of Offshore Wind Energy Resource in the United States of America for the 21st Century. *Appl. Energy* **2020**, *262*, 114537. [CrossRef]
66. Reeve, D.E.; Chen, Y.; Pan, S.; Magar, V.; Simmonds, D.J.; Zacharioudaki, A. An Investigation of the Impacts of Climate Change on Wave Energy Generation: The Wave Hub, Cornwall, UK. *Renew. Energy* **2011**, *36*, 2404–2413. [CrossRef]
67. Vagiona, D. Environmental Performance Value of Projects: An Environmental Impact Assessment Tool. *Int. J. Sustain. Dev. Plan.* **2015**, *10*, 315–330. [CrossRef]
68. Wu, Y.; Tao, Y.; Zhang, B.; Wang, S.; Xu, C.; Zhou, J. A Decision Framework of Offshore Wind Power Station Site Selection Using a PROMETHEE Method under Intuitionistic Fuzzy Environment: A Case in China. *Ocean Coast. Manag.* **2020**, *184*, 105016. [CrossRef]
69. Shokatpour, M.H.; Alhuyi Nazari, M.; Haj Assad, M. El Renewable Energy Technology Selection for Iran by Using Multi Criteria Decision Making. In Proceedings of the 2022 Advances in Science and Engineering Technology International Conferences, ASET, Dubai, United Arab Emirates, 21–24 February 2022. [CrossRef]
70. Saaty, T.L. *The Analytic Hierarchy Process*; McGraw-Hill: New York, NY, USA, 1980.
71. Asadi, M.; Pourhossein, K. Wind Farm Site Selection Considering Turbulence Intensity. *Energy* **2021**, *236*, 121480. [CrossRef]
72. Chaouachi, A.; Covrig, C.F.; Ardelean, M. Multi-Criteria Selection of Offshore Wind Farms: Case Study for the Baltic States. *Energy Policy* **2017**, *103*, 179–192. [CrossRef]
73. Díaz, H.; Guedes Soares, C. A Novel Multi-Criteria Decision-Making Model to Evaluate Floating Wind Farm Locations. *Renew. Energy* **2022**, *185*, 431–454. [CrossRef]
74. Kim, G.; Jeong, M.H.; Jeon, S.B.; Lee, T.Y.; Oh, H.Y.; Park, C.S. Determination of Optimal Locations for Offshore Wind Farms Using the Analytical Hierarchy Process. *J. Coast. Res.* **2021**, *114*, 439–443. [CrossRef]
75. Zadeh, L.A. Fuzzy Sets. *Inf. Control.* **1965**, *8*, 338–353. [CrossRef]
76. Wu, B.; Yip, T.L.; Xie, L.; Wang, Y. A Fuzzy-MADM Based Approach for Site Selection of Offshore Wind Farm in Busy Waterways in China. *Ocean Eng.* **2018**, *168*, 121–132. [CrossRef]
77. Deveci, M.; Cali, U.; Kucuksari, S.; Erdogan, N. Interval Type-2 Fuzzy Sets Based Multi-Criteria Decision-Making Model for Offshore Wind Farm Development in Ireland. *Energy* **2020**, *198*, 117317. [CrossRef]
78. Gil-García, I.C.; Ramos-Escudero, A.; García-Cascales, M.S.; Dagher, H.; Molina-García, A. Fuzzy GIS-Based MCDM Solution for the Optimal Offshore Wind Site Selection: The Gulf of Maine Case. *Renew. Energy* **2021**, *183*, 130–147. [CrossRef]
79. Taoufik, M.; Fekri, A. GIS-Based Multi-Criteria Analysis of Offshore Wind Farm Development in Morocco. *Energy Convers. Manag. X* **2021**, *11*, 100103. [CrossRef]
80. Tercan, E.; Tapkın, S.; Latinopoulos, D.; Dereli, M.A.; Tsiropoulos, A.; Ak, M.F. A GIS-Based Multi-Criteria Model for Offshore Wind Energy Power Plants Site Selection in Both Sides of the Aegean Sea. *Environ. Monit. Assess.* **2020**, *192*, 652. [CrossRef] [PubMed]
81. Calabrese, A.; Costa, R.; Menichini, T. Using Fuzzy AHP to Manage Intellectual Capital Assets: An Application to the ICT Service Industry. *Expert Syst. Appl.* **2013**, *40*, 3747–3755. [CrossRef]
82. Zhou, X.; Cheng, L.; Li, M. Assessing and Mapping Maritime Transportation Risk Based on Spatial Fuzzy Multi-Criteria Decision Making: A Case Study in the South China Sea. *Ocean Eng.* **2020**, *208*, 107403. [CrossRef]
83. Yager, R.R. A Procedure for Ordering Fuzzy Subsets of the Unit Interval. *Inf. Sci.* **1981**, *24*, 143–161. [CrossRef]
84. Pearl, J. *Probabilistic Reasoning in Intelligent Systems: Networks of Plausible Inference*; Morgan Kaufmann Publishers: San Francisco, CA, USA, 1988.
85. Borunda, M.; Jaramillo, O.A.; Reyes, A.; Ibargüengoytia, P.H. Bayesian networks in renewable energy systems: A bibliographical survey. *Renew. Sustain. Energy Rev.* **2016**, *62*, 32–45. [CrossRef]

86. Hosseini, S.; Sarder, M.D. Development of a Bayesian network model for optimal site selection of electric vehicle charging station. *Int. J. Electr. Power Energy Syst.* **2019**, *105*, 110–122. [CrossRef]
87. Schoemaker, P.J.H. Scenario planning: A tool for strategic thinking. *Long Range Plan.* **1995**, *28*, 117. [CrossRef]

Disclaimer/Publisher's Note: The statements, opinions and data contained in all publications are solely those of the individual author(s) and contributor(s) and not of MDPI and/or the editor(s). MDPI and/or the editor(s) disclaim responsibility for any injury to people or property resulting from any ideas, methods, instructions or products referred to in the content.

Review

A Review of Wind Clustering Methods Based on the Wind Speed and Trend in Malaysia

Amar Azhar * and Huzaifa Hashim

Department of Civil Engineering, University of Malaya, Kuala Lumpur 50603, Malaysia
* Correspondence: kva190010@siswa.um.edu.my

Abstract: Wind mapping has played a significant role in the selection of wind harvesting areas and engineering objectives. This research aims to find the best clustering method to cluster the wind speed of Malaysia. The wind speed trend of Malaysia is affected by two major monsoons: the southwest and the northeast monsoon. The research found multiple, worldwide studies using various methods to accomplish the clustering of wind speed in multiple wind conditions. The methods used are the k-means method, Ward's method, hierarchical clustering, trend-based time series data clustering, and Anderberg hierarchical clustering. The clustering methods commonly used by the researchers are the k-means method and Ward's method. The k-means method has been a popular choice in the clustering of wind speed. Each research study has its objectives and variables to deal with. Consequently, the variables play a significant role in deciding which method is to be used in the studies. The k-means method shortened the clustering time. However, the calculation's relative error was higher than that of Ward's method. Therefore, in terms of accuracy, Ward's method was chosen because of its acceptance of multiple variables, its accuracy, and its acceptable calculation time. The method used in the research plays an important role in the result obtained. There are various aspects that the researcher needs to focus on to decide the best method to be used in predicting the result.

Keywords: climate change; wind speed; wind trend; clustering; Ward's method; k-means

1. Introduction

Wind clustering plays an important role in determining the various aspects of the research objective, such as energy, engineering, and public safety. Therefore, the usage of the relevant clustering method is basically determined by the objective of the study and the parameters involved in the study. The sensitivity of the data also plays an important role in determining the method of clustering. It is important for the researcher to have an expectation of what the result should be and will be so that the method can be used efficiently.

This paper focuses on a comparison of the clustering methods used by researchers in terms of wind speed clustering. The areas considered in this paper are in Malaysia, Qatar, France, Iran, Turkey, the United States, India, South Africa, Switzerland, and Columbia.

The winds in Malaysia are influenced by two monsoon seasons: the southwest monsoon from late May to September and the northeast monsoon of Peninsular Malaysia from November to March. The heavy rain to the east of Peninsular Malaysia and of western Sarawak is caused by the northeast monsoon, whereas the southwest brings drought to the nation [1].

Figure 1 shows the northeast monsoon storms; the east of Peninsular Malaysia and Mersing are the windiest areas of Peninsular Malaysia. Therefore, according to various wind energy potential research studies in Malaysia, Mersing is often the best location for wind farming [2].

Citation: Azhar, A.; Hashim, H. A Review of Wind Clustering Methods Based on the Wind Speed and Trend in Malaysia. *Energies* **2023**, *16*, 3388. https://doi.org/10.3390/en16083388

Academic Editors: Yan Bao, Guanghua He and Liang Sun

Received: 20 February 2023
Revised: 31 March 2023
Accepted: 3 April 2023
Published: 12 April 2023

Copyright: © 2023 by the authors. Licensee MDPI, Basel, Switzerland. This article is an open access article distributed under the terms and conditions of the Creative Commons Attribution (CC BY) license (https://creativecommons.org/licenses/by/4.0/).

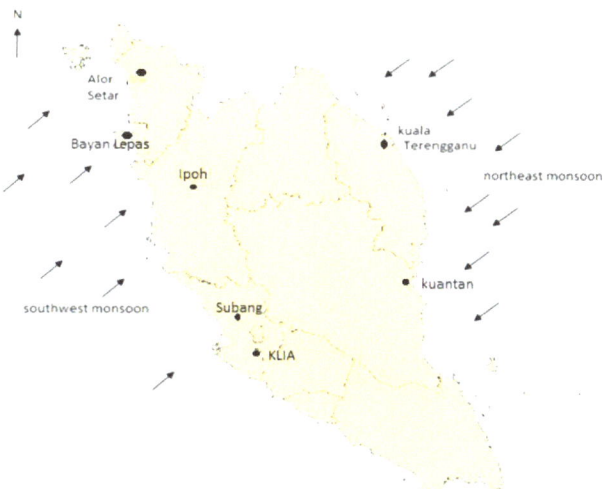

Figure 1. The direction of northeast and southwest monsoon in Peninsular Malaysia [1].

The monitoring of the wind speed trend is crucial for the prediction of future events or in seeing the continuity of the wind supply in certain areas. For example, the mapping conducted and adopted by the Malaysian Standard Code of Practice on Wind Loading for Building Structure uses mean wind speed. The standard is used by engineers in Malaysia, especially mechanical engineers and those involved with civil structures, to predict the wind speed in such areas as telecommunication antenna deployment. The mean wind speed usage may increase in a given year, as previously reported by Young in 2011. Figure 2 shows the recommendation by the Malaysian Standard on Basic Wind Speed with regard to a mean wind speed of 33.5 m/s [3].

Figure 2. Basic wind speed of Malaysia [3].

The global trends show that wind speed is increasing. Based on research in 2011, the global wind speed is increasing, indicating that extreme events are growing faster than the mean condition. The wind speed of most of the world's oceans has increased by at

least 0.25% to 0.5% per year. The strongest increasing trend was found in the southern hemisphere, and the northern hemisphere, especially the central North Pacific, shows a negative trend in wind speed. The wind speed increase in the central North Pacific was less than 0.25%, and some areas show a negative trend. As shown in Figure 3 below, the area surrounding Malaysia is also experiencing an increasing trend, especially in the southwest of Malaysia, where the southern Indian Ocean is located [4].

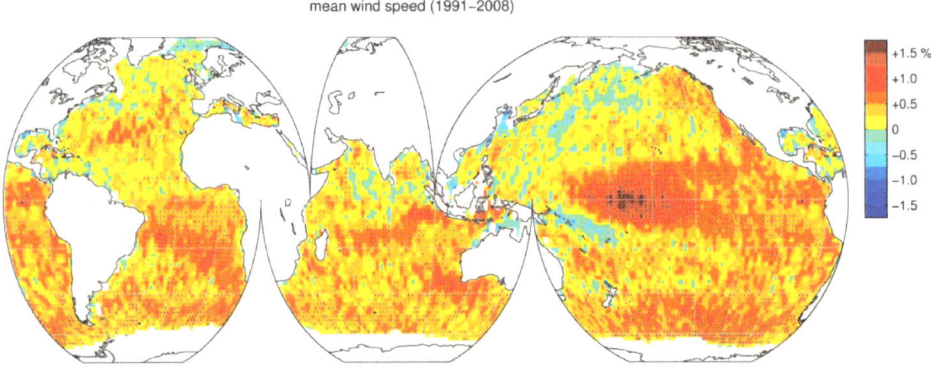

Figure 3. Mean wind speed (1991–2008) [4].

A study conducted in 2019 confirmed the above research by Young in 2011. As shown in Figure 4a, the research found that the global mean annual wind speed had increased for the previous ten years and that the pattern was increasing yearly. The Asian mean annual wind speed is also showing an increasing pattern. However, the wind speed in the Asian region began to increase earlier than the global speed. As shown in Figure 4b, the increase in wind speed in the Asian region started as early as 2002, whereas the global mean annual wind speed has been increasing since 2010. The research uses the diagnostic statistic for regression, which includes the goodness of fit, R^2, and the Pearson correlation coefficient, P. A Pearson correlation coefficient of less than 0.001 was considered be satisfactory in this study. However, the research found oscillation patterns that decreased the global wind speed; therefore, according to this research, wind energy production may decrease in the future [5].

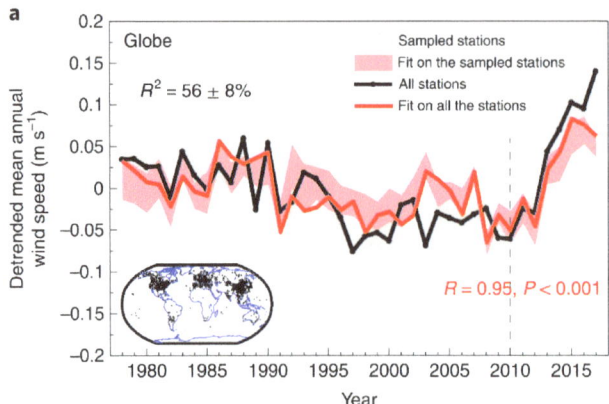

Figure 4. *Cont.*

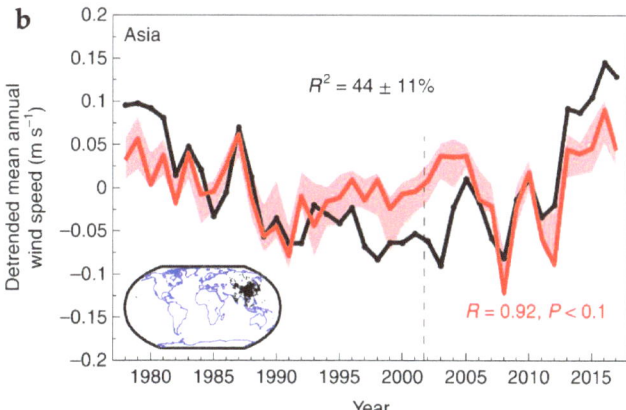

Figure 4. (a) Detrended mean annual global wind speed [5]. (b) Detrended mean annual wind speed in the Asian region [5].

In 2003, a study was conducted on the coastline of Peninsular Malaysia. The study focuses on analyzing the annual vector mean wind speed and direction according to two seasons, i.e., the northeast and southwest monsoons. The wind direction was northeast during the northeast monsoon season and southwest during the southwest monsoon season [6].

Research conducted in 2015 by Kok et al. found that the dynamics of the wind stress system had an important influence on the physical characteristics of the sea. The study used a wind stress curl to examine the mechanism responsible for the formation of the thermal front during both Malaysian monsoon seasons [7].

The positive and negative values of the wind stress curl cause cyclonic and anticyclonic motion in the northern hemisphere. This action causes divergence of the convergence in the surface layer of seawater. Therefore, the cooler or warmer water from the deep rises and replaces the diverging or converging water. This results in the upwelling and downwelling of the seawater. This upwelling is caused by the wind, which makes the water close to shore cooler [8].

Therefore, with regard to all of the wind characteristics mentioned, there is a need for wind trend monitoring and clustering, especially in Malaysia. Factors such as global warming have increased the temperature of the sea, causing the fluctuation in the global wind speed [9]. When wind standards commenced in 2002 in Malaysia, the need for wind clustering was foreseen; wind clustering can increase the accuracy of wind mapping and wind forecasting in Malaysia.

It is important for engineers and wind experts to able to see the wind trend and clustering according to objectives such as those considering area or demand. Each objective can show different results, which also depend on the method of clustering used. This paper aims to investigate the best method to cluster the wind trend. The specific objective is to determine the best method to cluster the wind trend in relation to the Peninsular Malaysia and Borneo regions.

2. Methodology of Wind Speed Clustering

2.1. Wind Speed Trend Observation

Wind speed observation has been conducted by the researchers based on various objectives. The research which uses the method of observation of wind speed trends is that of the wind energy researchers. Wind energy research requires wind trend observations to ensure the continuity of the wind supply that powers the wind harvesting equipment.

Research to evaluate the wind energy potential in Peninsular Malaysia was conducted from 2007 to 2009 by Masseran et al. at 10 wind stations. The research, which focuses on

the wind speed persistence in Peninsular Malaysia, is based on hourly data. The research found that for Peninsular Malaysia, the hourly wind speed for the wind station exhibits stationarity state. The smallest hourly wind speed observed at the Chuping station and Mersing station showed its suitability for the generation of energy due to its hourly wind trend. Therefore, the research shows the importance of wind trend observation in wind energy research methodology. Figure 5 below shows the wind speed trend for one week at Ipoh wind station, Perak [10].

Wind speed plot for Ipoh station

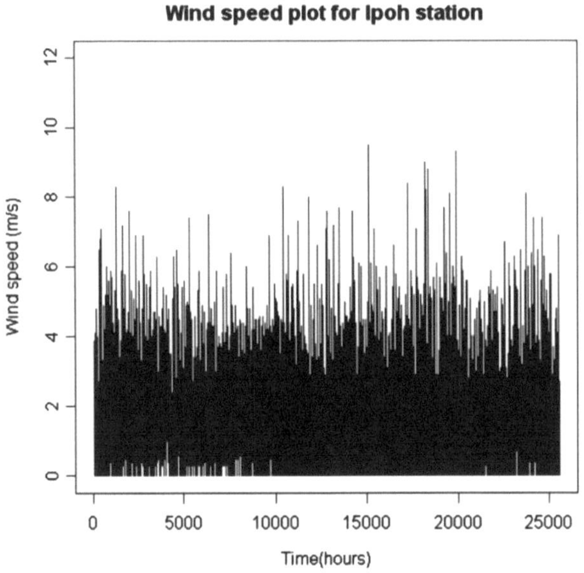

Figure 5. Wind speed trend at Ipoh wind station, Perak [10].

The wind trend also uses research conducted in Qatar by Aboobacker in 2021. The research uses monthly mean data to simulate the trend of the wind speed and to further estimate the wind power produced in the area [11]. The research focuses on the wind around the Arabian gulf coast and focuses on the Qatar peninsula. The research found that the highest wind speed was located in offshore Ruwais. Offshore Ruwais was found to be the windiest location and to have the highest mean wind speed. Table 1 below shows the wind speed statistics at the research locations from 1979 to 2018 [11].

Table 1. Wind speed statistics at onshore and offshore locations from 1979 to 2018 [11].

Region	Locations	Geographical Co-Ordinates		Wind Speed (m/s)			% of Exploitable Wind Speed
		Longitude (° E)	Latitude (° N)	Maximum	Mean	Standard Deviation	
Onshore	Mesaieed	51.5828	25.0444	15.4	4.6	2.3	73.9
	Al Khor	51.4394	25.7534	16.2	5.1	2.6	77.1
	Al Ruwais	51.2202	26.0690	15.9	4.9	2.5	74.7
	Dukhan	50.8398	25.3355	15.7	4.9	2.4	77.5
Offshore	Doha	51.7970	25.2755	15.5	5.1	2.5	78.5
	Ras Laffan	51.6146	26.0131	16.5	5.2	2.7	76.5
	Al Ruwais	51.2992	26.2822	16.9	5.5	2.8	78.4
	Dukhan	50.7251	25.4767	16.1	5.0	2.4	78.2

The trend observations were also conducted by research which showed that there were similarities in the onshore and offshore wind trends, as shown in Figure 6. However, the two windiest stations were Ras Laffan and Ruwais. The research using mean wind speed trend observations was similar to the current research in terms of the finding of the strongest wind recorded in the area. Therefore, the method of observation by trend was applicable in finding the windiest area or the area with the strongest wind.

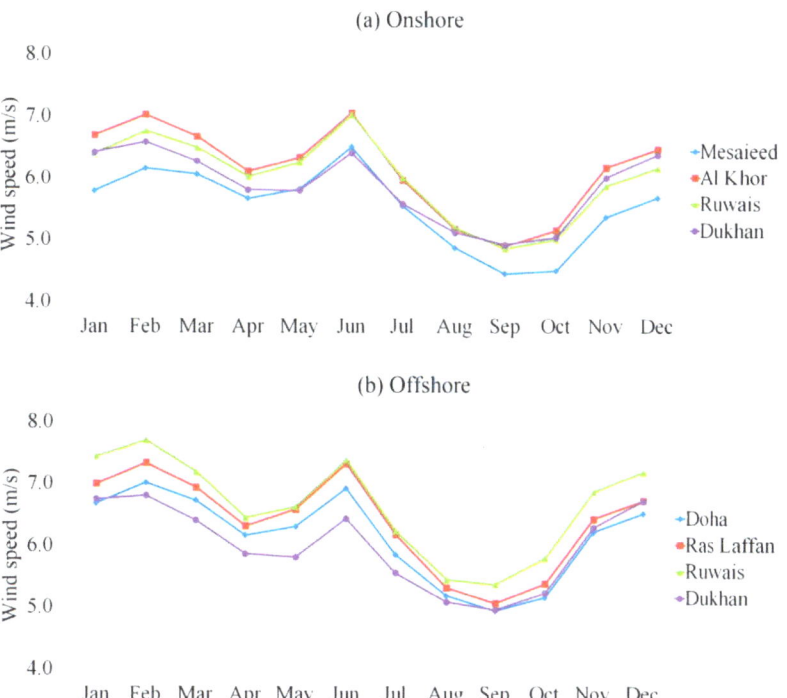

Figure 6. Monthly mean wind speeds at (**a**) onshore and (**b**) offshore locations of Qatar at a height of 90 m from 1979 to 2018 [11].

In 2012, Tiang and Ishak studied the wind speed at the measurement site of Bayan Lepas, Pulau Pinang, from January to December 2008. The study used wind trend observation to assess the potential wind energy in Pulau Pinang. By observing the trend of the wind speed, the researchers were able to find the windiest period in Pulau Pinang. Based on the findings, the maximum wind speed in Pulau Pinang was achieved in September, and the slowest was recorded in November. Using the trend observation, the researchers were able to determine the months in Pulau Pinang that were the windiest and had the highest wind speed; these were May, July, and September. The causes of the higher wind speed period were the southwest monsoon season and the geographical location of Pulau Pinang. Figure 7 below shows the monthly mean wind speed trend in 2008 from July to October [12].

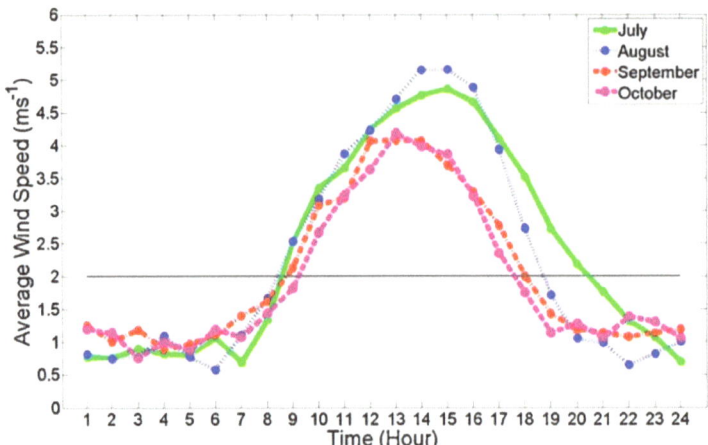

Figure 7. Monthly mean hourly wind speed in 2008 from July to October [12].

However, in terms of the engineering purposes, the wind trend observation focused more on the sudden spike in wind speed and the highest wind speed recorded in the research area. The research conducted by Shanmugasundaram et al. in 1998 was based on the tropical cyclone wind condition which occurred in June and December 1996. The research came out with a wind trend observation of the cyclone which indicated the highest mean and maximum wind speed recorded during the event. The wind speed trend observation helped the researchers to locate the maximum wind speed during the event and to calculate the damping ration increase for the 52 m steel lattice tower. Figure 8 below shows the mean and maximum wind speeds during the cyclone of the year 1996 [13].

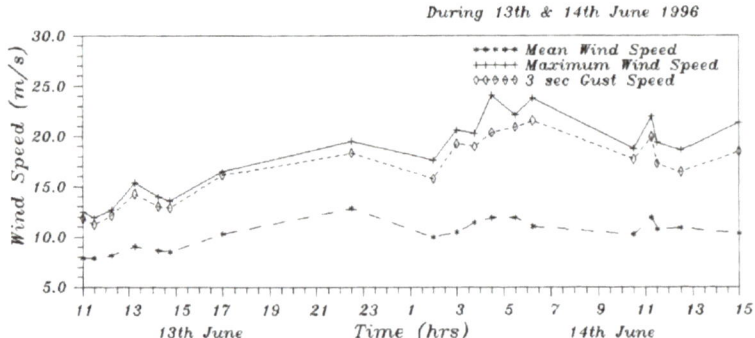

Figure 8. Mean and maximum wind speeds during cyclone [13].

The research which was based on the air pollution also used the wind trend analysis to simulate the severity of the pollution affecting the area. The direction and speed of the wind plays an important role in the air pollution effect. In 2016, Sokolov et al. conducted a cluster analysis of the atmospheric dynamics and pollution transport in the coastal area of industrialized Dunkerque in northern France. The research aimed to determine the trajectories in the context of pollution transport. The trajectories were based on the largest and most dispersed areas of low wind speeds, which make the pollution worse. The data of this research were based on the meteorological data of the wind speed and its direction and pollution measurements. The wind trend observation was visualized based on the wind rose. The wind rose modeling was successful in showing the trend in terms of the

direction and the wind speed at Maregraph station. Figure 9 below shows the wind rose modeling for Maregraph station from 1st May to 1st October 2006 [14].

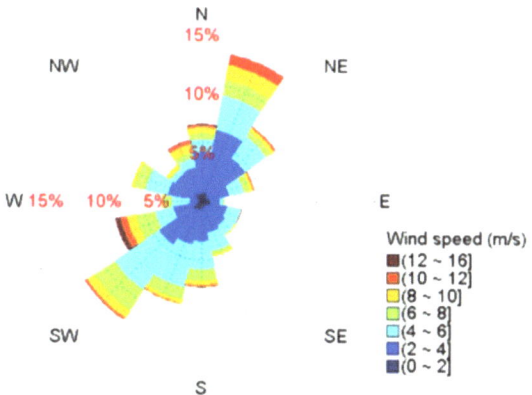

Figure 9. Wind rose for modeling period 1 May to 1 October 2006 [14].

Therefore, the wind trend observation requirement is based on the objective of the research. The wind trend observation can assist with multiple factors and can contribute to the objective of the research. However, wind trend observation for a longer period may require grouping or clustering to ease the analysis and to localize the wind trend according to the area.

2.2. Clustering Wind Speed

2.2.1. Linkage–Ward Clustering Method

The probabilistic wind speed clustering was used in the study cases at Khaaf, Iran, in 2018 [15]. Azizi et al., reported using the Linkage–Ward clustering method to cluster the wind speed in the area. The research reported that the usage of the Ward clustering method was higher in accuracy compared to the k-means method. The Ward method, however, was more complex than the k-means method. For two years, the study used the measured wind speed time of 60 min in the wind stations around Binalood, Iran. The wind stations vary in height, soil, and distance to residential areas. The focus of the study was to select the proper site to install the wind turbine in Binalood. Therefore, the study focuses on the windiest area, which can be correlated with the current study. Although the study also used the Linkage–Ward clustering method instead of k-means, the Linkage–Ward clustering method required even more computational effort to solve.

The research found that the Linkage–Ward clustering method was the most common and accurate for use in the study. The method calculated the dissimilarity between clusters based on the centroid of the cluster, as shown in (1)

$$d_{i+j,k} = ad_{ik} + ad_{jk} + bd_{ij} + c\left|d_{ik} - d_{jk}\right|, \qquad (1)$$

where d_{ik}, d_{jk}, d_{ij} are the pairwise distances between the clusters i and k, j and k, and i and j. i, j, k are the indexes of the clusters. n_i, n_j, n_k are the numbers of members within clusters i, j, and k, respectively.

$$a = \frac{n_i + n_k}{n_i + n_j + n_k}, \quad b = \frac{n_j + n_k}{n_i + n_j + n_k} \qquad (2)$$

where a and b are defined as (2), and $c = 0$ in the Linkage–Ward clustering method. a and b are the parameters, which depend on the cluster size to determine the clustering algorithm, with a distance between clusters of d_{ij}.

The clusters which have the lowest increase in distance between the cluster centroids (1) are combined. The Ward method uses the objective function in the sum of the squares from the points to the centroids of the clusters. Figure 10 below shows the step-by-step algorithm of Linkage–Ward clustering.

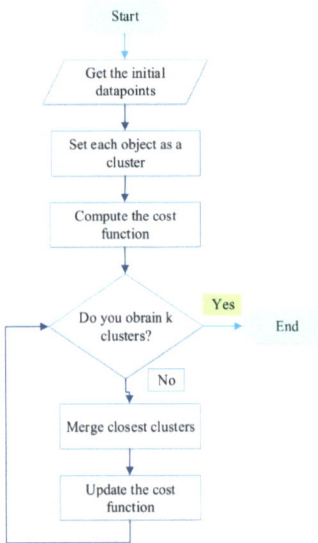

Figure 10. Linkage–Ward clustering step-by-step algorithm [15].

The calculation above will result in the lowest increase in the cost function of (1) and in the combination. The method uses the objective function in the sum of the squares from the points to the centroids of the clusters. Figure 11 below shows the average wind speed value sample at a 40 m height with 10 min intervals in the study area. The color lines indicate 50 days chosen randomly by the researchers [15].

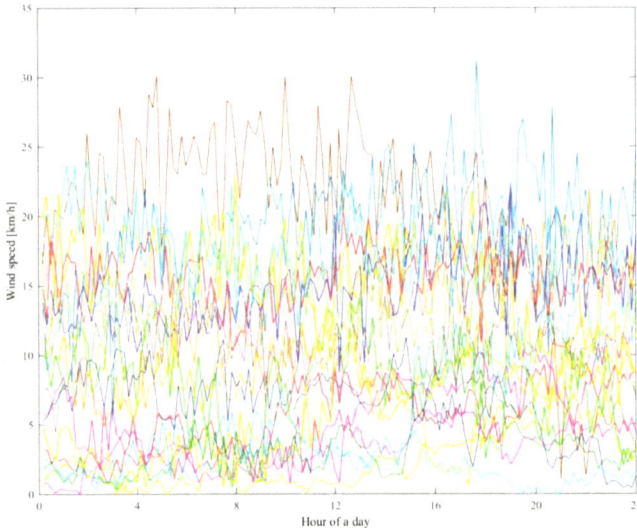

Figure 11. Measured wind speed for 50 days at Khaaf, Iran [15].

The researchers found the centroid of the cluster where the study was able to find the mean of the wind speed earlier in the research. This is a reverse method to find the centroid of the cluster and may affect the result. Figure 12 below shows the cluster centers of the measured wind speed.

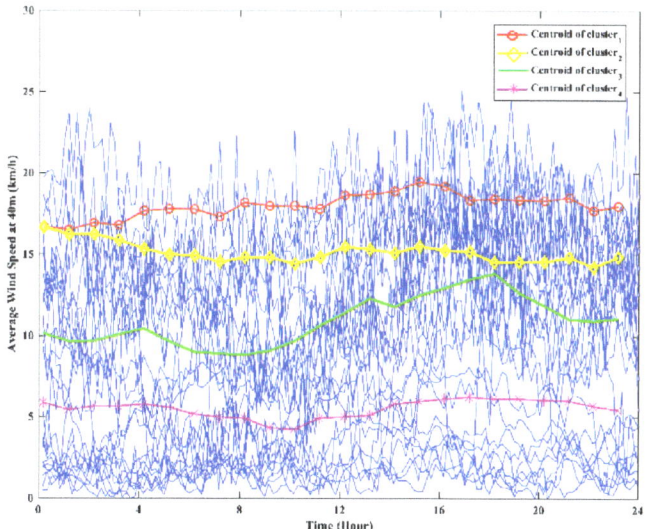

Figure 12. Cluster centers versus dataset [15].

The number of clusters was chosen by calculating the error of the cluster's centroid and its member. As expected, a small number of clusters brings out the dissimilar object group. The optimal number of clusters is important to ensure the effectiveness and the accuracy of the data. The Euclidian error between each cluster is calculated as in (3).

$$error = \sum_{i=1}^{N_{Cluster}} \sum_{j=1}^{n_j} |x_j - c_i| \qquad (3)$$

where $N_{Cluster}$ is the number of clusters, n_j is the number of members within cluster j, respectively, x is each observation in the dataset, and c_i is the centroid of cluster i.

The error calculation found that the minimum error obtained for this research was four clusters, as the calculation showed that only an 8% relative error was found. Therefore, the research used four clusters as the basis of the clustering for the dataset. Figure 13 below shows the error calculation result in determining the number of clusters.

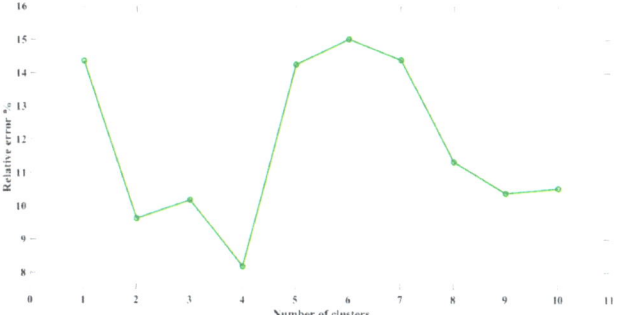

Figure 13. The error for different numbers of clusters [15].

Azizi et al. found that from the four clusters created, cluster 2 had the higher probability compared to the other clusters, at 38%. The higher probability occurrence suggests that cluster 2 is more suitable for wind farming. Figure 14 below shows the probability of occurrence of each cluster.

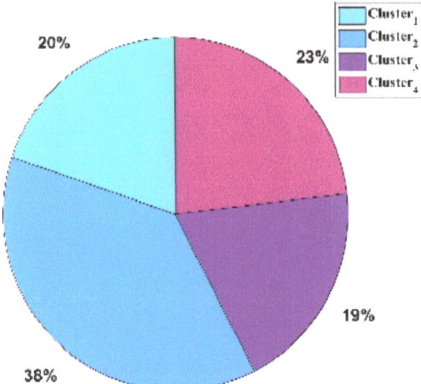

Figure 14. Probability of occurrence of each cluster [15].

2.2.2. k-Means Approach for Wind Clustering

The annual wind speed patterns can be grouped when the study area is the same. Yesilbudak et al. conducted a clustering analysis of multidimensional wind speeds for 75 provinces in Turkey. The method used in the clustering was the k-means approach. In this research, the silhouette coefficient was used to determine the effectiveness of the distance measure. The analysis found that the prominent cities in terms of average wind speed were Canakkale and Mardin, located in cluster 4, where the mean cluster of silhouettes was 0.5224. On the other hand, cluster 1 contained Duzee, Amasya, and Siirt, which were determined to be poorly matched areas with the silhouette coefficients of 0.7294, 0.7198, and 0.7111. Figure 15 below shows the silhouette coefficients for k = 5 and the square Euclidean distance measure result [16].

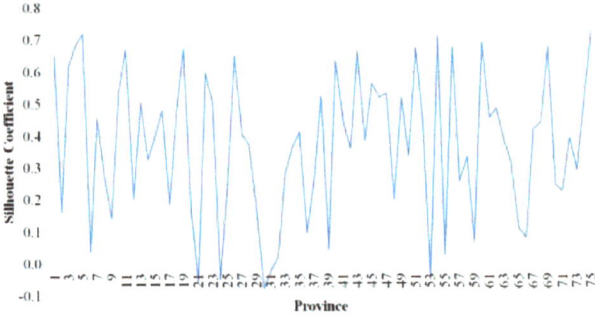

Figure 15. Silhouette coefficients for k = 5 and square Euclidean distance measure result [16].

In this research, the study mentioned k-means as one of the portioning methods in the literature. The k-means algorithm assumes that D is the dataset that contains n observations and k is the number of clusters. The k-means calculated the dissimilarity between each pair observation differently according to the distance measures. Four types of distance measures were used: squared Euclidean, city-block, cosine, and Pearson. Figure 16 below shows the k-means algorithm used in the study.

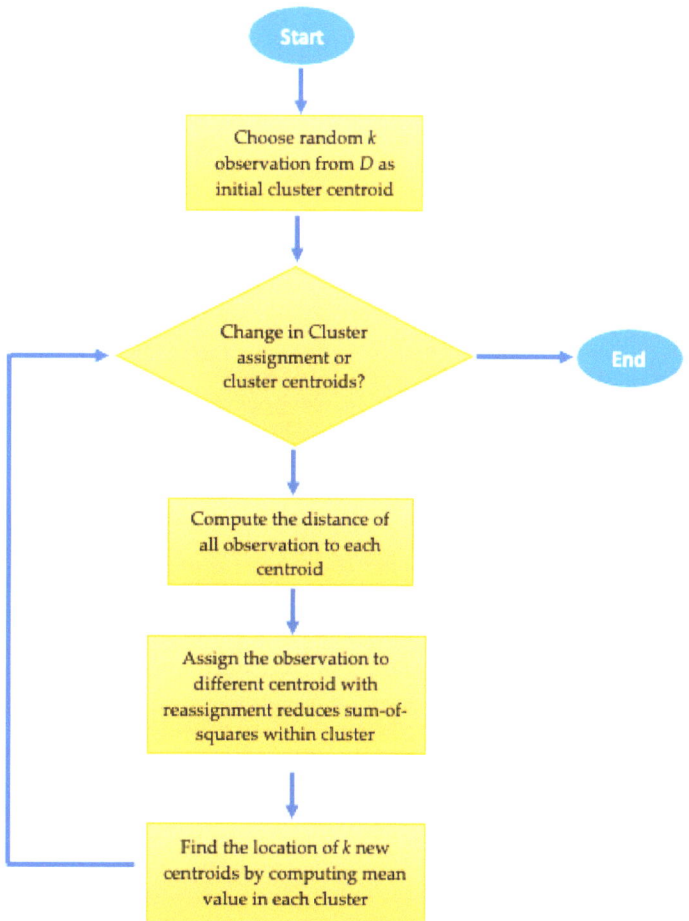

Figure 16. k-means algorithm method flowchart.

To determine the best distance measure, the silhouette coefficient varying between −1 and +1 was used for measuring the observation assigned to the clusters. The accuracy was defined by the silhouette coefficient closer to 1, which indicated that the observation belonged to its cluster. The silhouette was defined as in (4) below.

$$s(y_i) = \frac{b(y_i) - a(y_i)}{max\{a(y_i), b(y_i)\}} \quad (4)$$

where $a(y_i)$ is the average dissimilarity of y_i and the element of ($\in$) S_k to all other $y_j \in S_k$, and $b(y_i)$ is the minimum average of dissimilarity of $y_i \in S_k$ to all other $y_j \in S_l$.

As shown in Figure 17, the study plots the annual wind speed data using star glyph plots. The plots shown in Figure 17 show the wind pattern of the 75 areas around Turkey. The analysis by the k-means algorithm with the silhouette coefficient gives a stronger clustering solution. The research found that using the square Euclidean distance measure gives a more accurate clustering result compared to the other three distance measuring methods. Therefore, the clustering result was obtained using the square Euclidean distance measure, as shown in Table 2 below.

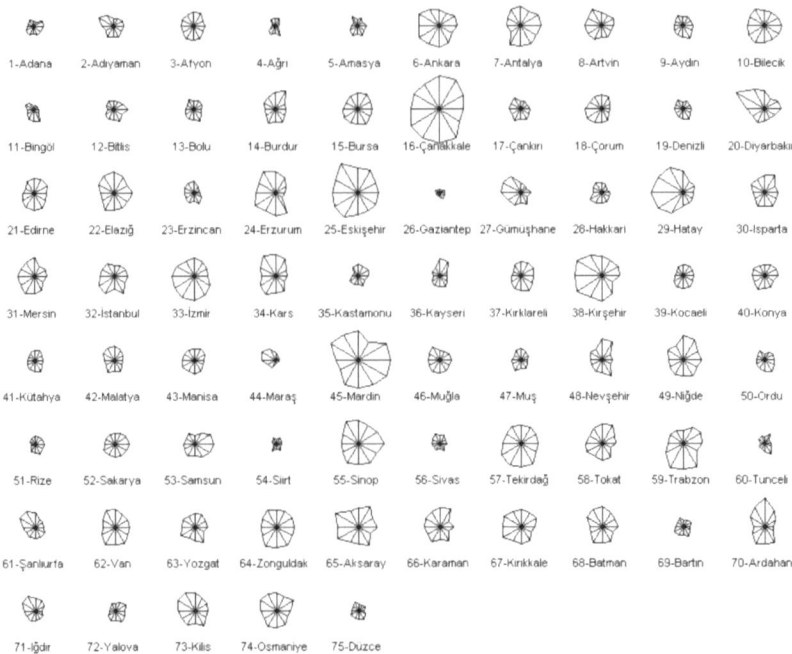

Figure 17. The star glyph plots created for visualizing multidimensional wind speed data [16].

Table 2. The province categorized into each cluster by the k-means approach [16].

Cluster Name	Cluster Observations
Cluster 1	1, 4, 5, 9, 11, 12, 13, 17, 19, 23, 26, 28, 35, 36, 41, 44, 47, 50, 51, 54, 56, 60, 69, 72, 75
Cluster 2	7, 10, 15, 20, 21, 22, 30, 31, 32, 34, 49, 53, 58, 62, 64, 66, 67, 68, 70, 73, 74
Cluster 3	2, 3, 8, 14, 18, 27, 37, 39, 40, 42, 43, 46, 48, 52, 61, 63, 71
Cluster 4	16, 45
Cluster 5	6, 24, 25, 29, 33, 38, 55, 57, 59, 65

Time series clustering has been widely used in predicting wind speed. For example, Kusiak et al. conducted wind speed clustering to predict the power output generation based on the wind speed. The researchers' study was based on the long- and short-term prediction of power using the k-nearest neighbor (k-NN) algorithm [17].

In this research, multiple parameters were considered during clustering calculation. The parameters also made the clustering much more detailed and precise. Therefore, a clustering method that can cater for bigger variables has to be used for the clustering exercise to be successful. Table 3 below shows the list of parameters used in the research by Kusiak et al.

However, the current wind speed data were unavailable during the study. Therefore, the prediction of the power generated from the wind speed was not validated [17].

In 2012, Andrew Clifton demonstrated the usage of k-means clustering to identify the relationship between the wind at turbine height and climate oscillation. The study used fourteen years of data from an 80 m tower at the National Wind Technology Center (NWTC) in Colorado. During the study, the k-means method of clustering identified four dominant wind flows in the area. The study first identifies the frequency of the wind direction. However, for the frequency study, the data are limited to the wind speed of 3.5 m/s and

grouped into 5° and 1 m/s bins. The contours show the relative frequency in each bin on a linear scale. Figure 18 below shows the wind frequency visualized in contours.

Table 3. List of parameters selected for wind speed estimation [17].

Parameter Type	Parameter Name	Abbreviation	Symbol	Unit
Non-controllable	Wind speed	WS	v	m/s
Controllable	Blade pitch angle	BPA	x_1	
	Generator torque	GT	x_2	Nm
Performance	Power output	PO	y_1	kW
	Rotor speed	RS	y_2	rpm

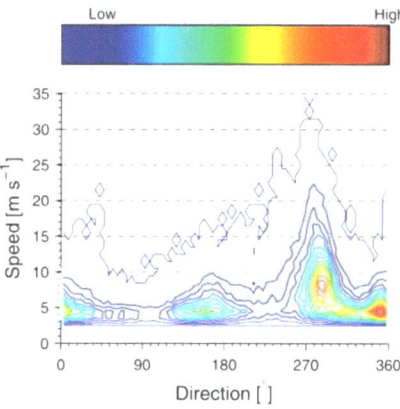

Figure 18. Frequency of wind at 80 m binned by wind speed and direction [18].

The researcher applied the k-means clustering approach to zonal and meridional wind speeds. The k-means clustering splits N data points into k clusters and assumes that the data belong to the nearest mean value. The researcher repeated the clustering 100 times using a random initial centroid and generated an optimum set of centroids. The research used the function form of the "Statistics Toolbox" in the software MATLAB R2010b to generate the k-means analysis. Thereby, four dominant flows were found: the weak northerly (N), weak southerly (S), weak westerly (W(L)), and strong westerly flows. The clustering of the flows is shown in Figure 19 below [18].

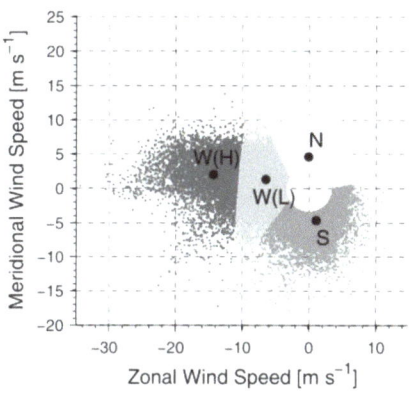

Figure 19. Optimal wind clusters at 80 m at the NWTC near Boulder [18].

The optimum number of clusters was obtained by Andrew Clifton's research using the Bayesian information criterion (BIC) method. The BIC method increased the number of k to a point where k would not give a meaningful quality to the result. The method performs well in two-dimensional datasets, especially when using a machine learning application such as MATLAB [19]. Figure 20 below shows the variation of the normalized BIC value with the number of clusters, and the result shows that optimum number of k is 4.

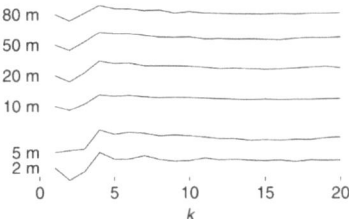

Figure 20. Variation of normalized BIC value with number of clusters when M2 meridional and zonal winds are grouped into k clusters at each height [18].

2.2.3. Non-Parametric Approach Hierarchical Clustering

Guldal et al. used hierarchical clustering algorithms to cluster the wind speed and blow number, a parameter which causes evaporation in Lake Egirdir, Turkey. The research used a non-parametric approach of the hierarchical clustering algorithm where the monthly evaporation losses and the mean wind speeds with the blow number were clustered. The clustering method was determined by the mutual neighbor distance (MND) algorithm. Figure 21a shows the pattern labelled A, B, C, D, E, F, and G, which falls into three clusters. The clustering can be further refined using a single-link algorithm, as shown in Figure 21b [20].

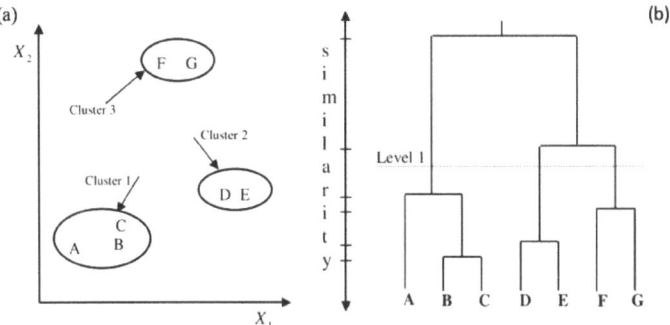

Figure 21. The two-dimensional dataset (a) and dendrogram obtained using single-link algorithm (b) [20].

Figure 21 shows the hierarchical clustering algorithm in a two-dimensional dataset. Figure 21a shows that there are seven observations, labelled as A, B, C, D, E, F, and G, in three clusters. Therefore, in Figure 21b, the dendrogram shows the grouping of seven patterns and the similarity levels of the observations. Figure 21b shows that the clustering can be broken into multiple levels. For example, level 1 comprises three clusters, (A, B and C), (D and E), and (F and G) [20].

The mutual neighbor distance (MND) used by this study is described in Figure 22 and by MND Equation (5) below;

$$MND(x_i, x_j) = NN(x_i, x_j) + NN(x_j, x_i) \qquad (5)$$

where $NN(x_i, x_j)$ is the neighbor number of x_j with respect to x_i. Figure 22 shows the example of MND. The neighbor nearest to A is B, and B is the nearest neighbor of A. Therefore, $NN(A,B) = NN(B,A) = 1$. The MND between A and B become 2 according to Equation (5) above. The $NN(B,C) = 1$, and the $NN(C,B) = 2$. Therefore, the $MND(B,C) = 3$.

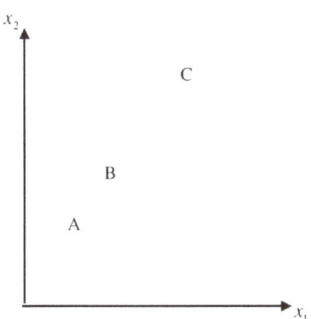

Figure 22. A and B are more similar than A and C.

The result from the above method shows both the similarity (S) levels (l) of S6 (l6) and S8 (l8) and the strong relation of the evaporation rate, R^2 ($R^2 = 0.29$ for wind speed change and evaporation rate), ($R^2 = 0.85$ for wind blow number and evaporation rate), for June, July, August, and September. The strongest relationship is the clustering at l = 6 (S6), as shown in Figure 23a; the detail of the similarity level S6 (l6) clusters analysis is shown in Figure 23b, where the coefficient of the evaporation rate is 0.96. Therefore, the clustering should determine different operation levels to make efficient operating decisions and accurate predictions. Furthermore, this prediction should produce scientific meaning by representing the actual object in the best way [20].

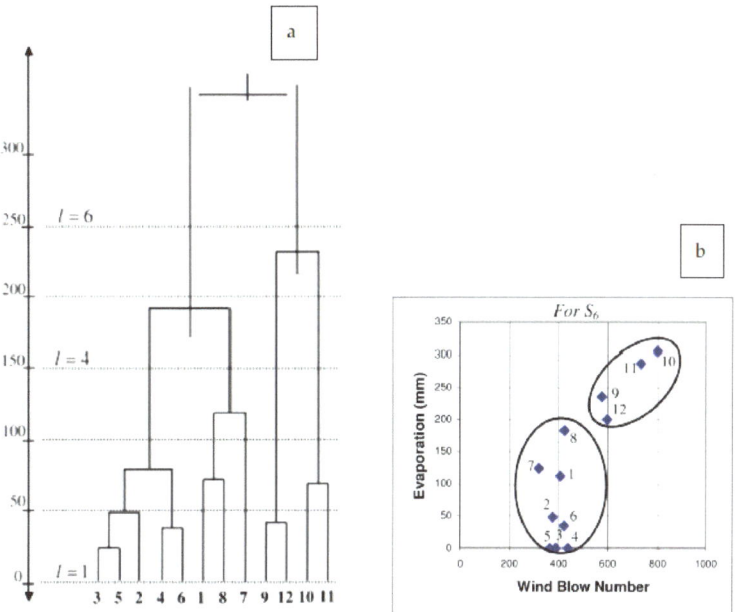

Figure 23. The dendrogram depends on the hierarchical single linkage for the second application (**a**) and detail of similarity level S6 (l6) cluster analysis (**b**) [20].

However, the research of Guldal et al. does not discuss the relative error or comparison between methods since the research only uses the non-parametric approach.

2.2.4. Trend-Based Time Series Data Clustering Using Statistical Model

The wind prediction method has been studied and revised with multiple hybrid methods to simplify and increase the accuracy of the algorithm. Kushwah et al. studied wind forecasting by using a time series. Wind components such as seasonal trends can be monitored in the time series application. In this research, the clustering method was based on the seasonal trend. As shown in Figure 24 below, the proposed model for wind speed prediction uses the trend as the major component during the study [21].

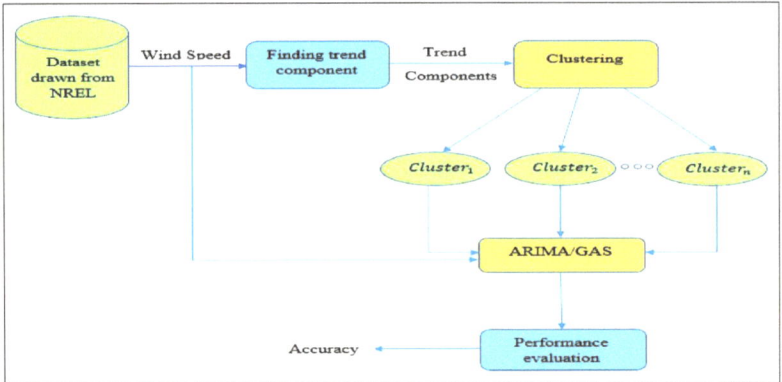

Figure 24. Proposed model for wind speed prediction [21].

The study used standard deviation for data analysis. The result from the standard deviation analysis was then converted into a time series for clustering purposes. The wind prediction was evaluated in four models: the autoregressive integrated moving average (ARIMA), the generalized autoregressive score (GAS), a hybrid model of C-ARIMA, and a hybrid model of C-GAS. The finding was that both hybrid models performed better compared to the original model of ARIMA and GAS in terms of forecasting wind trends. Figure 25, in the left, middle, and right panels, shows the wind speed prediction using the GAS model for the first, second, and third clusters.

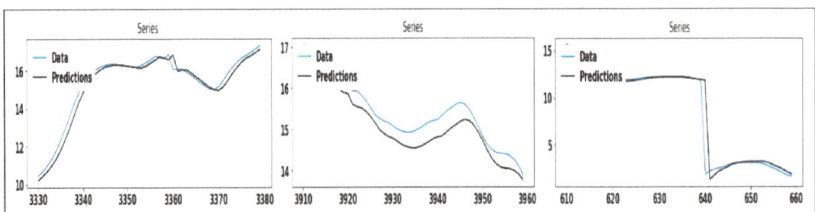

Figure 25. The wind speed prediction using the GAS model on dataset #1 [21].

The result also shows that the mean absolute error (MAE) and root mean square error (RMSE) for the hybrid models are lower than the original, as shown in Tables 4 and 5 below. The bolded numbers in the tables are the lowest error values obtained during the analysis [21].

Table 4. MAE and RMSE values using the ARIMA and clustered ARIMA models [21].

Dataset	ARIMA		CI-ARIMA		C2-ARIMA		C3-ARIMA	
	MAE	RMSE	MAE	RMSE	MAE	RMSE	MAE	RMSE
#1	7.346	8.649	**5.159**	**5.973**	5.940	6.937	5.570	6.326
#2	4.675	5.972	7.029	8.43	**4.191**	**5.206**	4.726	6.047
#3	2.968	4.281	**2.747**	**3.897**	3.821	5.021	6.156	6.964
#4	6.593	6.976	**6.374**	**6.757**	9.113	9.383	13.893	14.061
#5	**4.074**	**4.359**	10.319	11.136	6.344	6.685	7.404	7.697
#6	**5.558**	**7.124**	5.563	7.254	5.653	7.331	5.679	7.241
#7	2.796	3.291	3.257	3.833	**1.714**	**2.106**	92.738	123.995
#8	4.246	4.95	**3.362**	**4.049**	3.554	4.252	4.88	5.645
#9	**4.207**	**4.785**	11.621	12.062	9.483	10.01	9.358	9.877
#10	3.455	3.771	2.528	2.869	4.725	4.983	**2.294**	**2.591**
#11	2.188	2.743	2.144	2.699	**2.048**	**2.609**	6.471	6.956
#12	4.751	6.593	4.22	4.808	11.128	12.232	**3.917**	**4.949**

MAE: mean absolute error; RMSE: root mean square error; ARIMA: autoregressive integrated moving average.

Table 5. MAE and RMSE values using the GAS and clustered GAS models [21].

Dataset	GAS		CI-GAS		C2-GAS		C3-GAS	
	MAE	RMSE	MAE	RMSE	MAE	RMSE	MAE	RMSE
#1	5.017	5.767	**2.955**	**3.398**	6.324	7.377	6.454	7.342
#2	4.788	6.176	5.776	7.304	**4.51**	**5.672**	4.802	6.133
#3	3.003	3.82	13.468	15.224	**1.785**	**2.395**	2.887	3.356
#4	2.738	3.431	**1.794**	**2.449**	4.767	5.94	9.919	10.25
#5	**5.212**	**5.373**	6.928	7.107	8.11	8.579	7.266	7.559
#6	7.268	9.359	7.175	9.369	**5.94**	**7.771**	6.143	7.955
#7	1.851	2.599	3.155	3.47	**1.715**	**2.016**	5.188	5.402
#8	6.324	7.166	5.487	6.258	5.341	6.077	**5.029**	**5.777**
#9	7.068	8.019	5.52	6.553	**4.321**	**5.37**	7.387	8.227
#10	4.952	5.371	**2.112**	**2.426**	8.576	8.8	2.82	3.159
#11	2.474	3.164	2.988	3.468	**2.42**	**3.078**	8.334	8.76
#12	**4.378**	**5.951**	5.9	6.899	5.213	6.823	6.093	7.259

MAE: mean absolute error; RMSE: root mean square error; GAS: generalized autoregressive score. Bold numeric value of MAE and RMSE indicates that the prediction model corresponding to the column has the least prediction error and performed better on the Dataset representing that row.

The study above, however, did not reveal the result of the wind clustering and only reviewed the precision of both hybrid methods of wind forecasting.

In 2019, based on the Komsberg, South African area, research on the mean daily wind speed was conducted by Vuuren and Vermeulen. The study focuses on clustering the mean daily wind speed and comparing it with the customers' demands. The research then further analyzed the tariff to optimize the siting areas for wind energy facilities. The study used multiple clustering algorithms to cluster wind resource datasets. The algorithms used were k-means, partitioning around medoids, the clustering large application algorithm, agglomerative clustering, the divisive analysis algorithm, and fuzzy c-means clustering. The research also used the Euclidean distance and Pearson correlation of the distance measurement. The research used the standard deviation method to obtain the mean high wind

speed. Figure 26 shows the daily mean, median, and variance characteristics of the wind speed profiles for the REDZs for the 2013 period, using the standard deviation method [22].

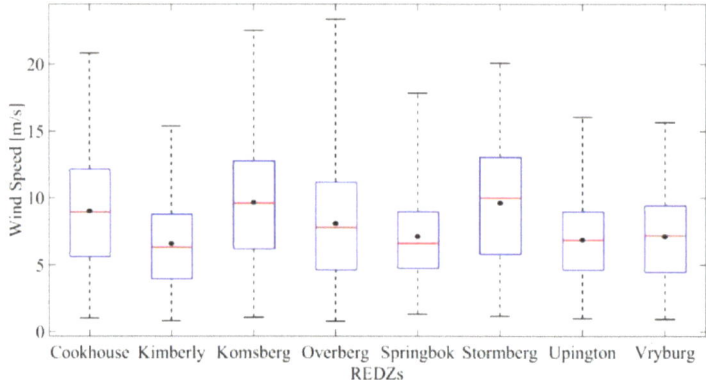

Figure 26. Boxplot showing the daily mean, median, and variance characteristics of the wind speed using the standard deviation method [22].

The research used three types of clustering methods. The clustering methods were the k-means algorithm, the partitioning around medoids algorithm (PAM), and the clustering large application algorithm (CLARA). The k-means clustering algorithm result showed a non-overlapping cluster for the Komsberg wind speed profile. Figure 27 below shows a 2D representation of the variables through principal component analysis.

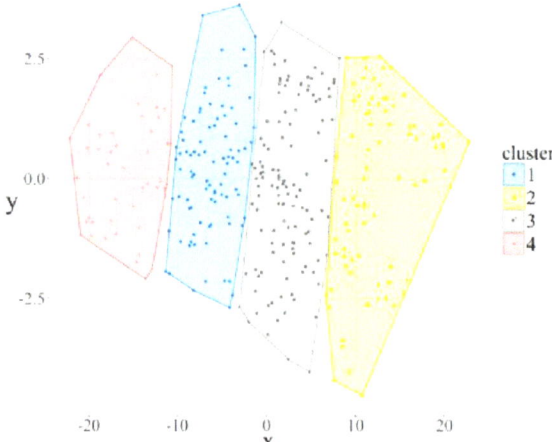

Figure 27. Non-overlapping clusters obtained with the k-means algorithm [22].

The research used a dendrogram to show the cluster assignment obtained by using the hierarchical agglomerative algorithm. Figure 28 below shows the clustering tree-like structure used to represent the four clusters assigned to the data based on the clustering method. Therefore, the mean wind speed can be visualized by the tree diagram and is easy to understand.

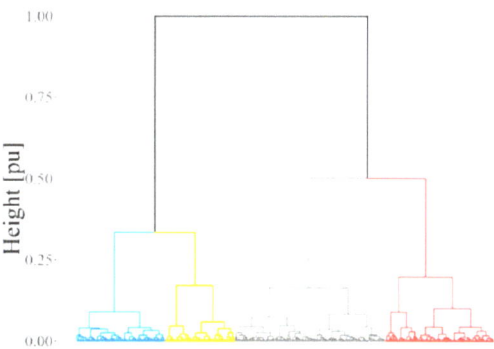

Figure 28. Dendrogram representation of the tree-like structure obtained with the hierarchical agglomerative algorithm [22].

Based on the clustering analysis, Table 6 below shows the validation result of the research. The result shows that the PAM and CLARA algorithm gave the best validation result. It was found that the CLARA algorithm reduced the algorithmic computing time of the large datasets without deceasing their accuracy. The CLARA algorithm also gave the highest silhouette coefficient. Therefore, it was concluded that CLARA algorithm was the most suitable method to use in this research.

Table 6. Validation result for the various clustering algorithms.

Validation Method	Partitioning Clustering Algorithms			Hierarchical Clustering Algorithms		Advance Algorithms
	K-Means	PAM	CLARA	Agglomerative Clustering	DIANA	Fuzzy C-Means
Silhouette coefficient	0.44	0.48	0.52	0.45	0.47	0.45
Number of incorrect cluster assignments	14	6	2	26	4	16
Calinski-Harabasz Index	861.731	851.731	826.047	794.786	854.381	829.219
Average distance within clusters	4.436	4.423	4.516	4.584	4.446	4.433
Dunn index	0.0569	0.0340	0.0411	0.0455	0.0420	0.0410

2.2.5. Anderberg Hierarchical Clustering Method

In 1996, Kaufmann et al. used the hierarchical clustering method in research in which the wind speed was an absolute value with vector differences at the station. The research took place for a duration of one year in the city of Basel. The period reflected the diurnal and seasonal airflow variation in the complex terrain. The study analyzes the normalized hourly mean of the wind fields. The distances measured for the study were defined as the mean absolute values of the vector differences at all the stations involved [23]. The study is comparable to the study of Gassmann et al., in which they used Ward's clustering method with distances of Euclidean measurement [24].

However, the method was found to be unsuitable for use in the study. Therefore, the study used the complete linkage method (Anderberg), which tended to build a group of similar size but focused on the ranking of the distances. Table 7 below summarizes the

result obtained in the study of Kaufmann et al., where 15 clusters were found based on the criteria given [23].

Table 7. Summary of the 15 clusters obtained by the complete linkage clustering method [23].

Cluster Number	Count	Relative Frequency	Mean Wind Speed (ms^{-1})	Std. Dev. (ms^{-1})
1	127	1.4%	1.7	0.60
2	166	1.9%	1.5	0.41
3	419	4.8%	2.1	0.71
4	30	0.3%	2.0	0.74
5	1692	19.3%	3.8	1.76
6	751	8.5%	2.8	1.19
7	637	7.3%	2.3	0.82
8	1643	18.7%	2.5	0.83
9	376	4.3%	1.9	0.61
10	580	6.6%	3.2	2.10
11	176	2.0%	1.3	0.27
12	407	4.6%	2.3	1.41
13	159	1.8%	1.4	0.39
14	1050	12.0%	1.9	0.59
15	571	6.5%	2.0	0.67
Total	8784	100.0%	2.6	1.39

The study found that 15 clusters could be produced based on the analysis using complete linkage clustering. A clear diurnal variation of wind patterns was observed, and it fit with the physical mechanism of the mountain valley wind and the characteristics of the sample of the cluster for normalized wind vectors obtained during the study, as in Figure 29 below. The research, however, did not discuss the error analysis of the method used [23].

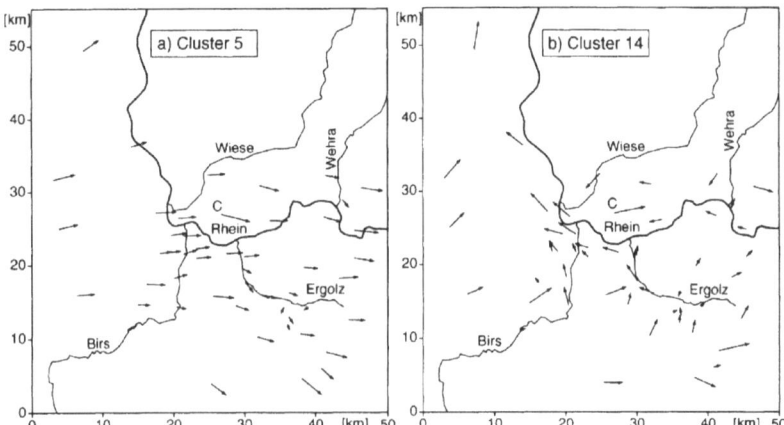

Figure 29. Cluster averages of normalized wind vectors at all measurement sites for (**a**) cluster 5 and (**b**) cluster 14. "C" labels the station on the TV tower at St. Chrischona [23].

2.2.6. Other Methods of Data Clustering

Angosto et al. conducted a wind clustering analysis to predict atmospheric pollution. The research found five different wind patterns by using a two-step clustering analysis in the city of Cartagena. The analysis clustered the wind direction into five clusters. For example, the first cluster found that 6.5% of the cases of wind direction were north-northwest and north. The second cluster had wind of a south-southwest and south direction, which comprised 24.7% of the data. The method used in this research was a two-step clustering analysis procedure that used the hierarchical (average linkage) and non-hierarchical (k-means) methods [25].

There are other clustering algorithms, such as the density-based spatial clustering of application with noise (DBSCAN) and the autoregressive integrated moving average (ARIMA). Dokuz et al. used both the DBSCAN and the ARIMA algorithms in their research on wind speed forecasting. The study found that using both methods provided a better performance than using a single method. In addition, the hybrid method proved that the root mean square error (RMSE) decreased up to 20% [26].

3. Recommendation and Conclusions

As mentioned in the above topics, there are many methods of clustering used to cluster wind speed. The non-parametric hierarchical clustering using the mutual neighbor distance algorithm shows a complex method of clustering and an acceptable result. The method showed an efficient operating decision and made accurate predictions during research [20].

The trend-based time series clustering shows that the method produces excellent accuracy. Even though the research focuses on forecasting the wind speed, the study shows that the wind speed can be clustered according to its trend. This was shown in the research of Kushwah et al. for the yearly trend. Therefore, the trend can be predicted as it follows a seasonal pattern, and the application of this research is good for research with a localized wind speed trend prediction. The clustering using the trend-based method was successfully shown in the research of Vuuren et al., where the researchers successfully clustered the mean daily wind speeds for the high demand season using the clustering large application algorithm (CLARA).

However, there are two main methods that the wind clustering researcher usually uses: the k-means and the Ward methods. Both methods are based on the k-value to determine the partition size of the cluster. The cluster size is important to the researcher when determining the number of desired clusters according to the research objective.

For the k-means method, the algorithm gives no guidance for the numbers of k. However, Ward's method gives some partition sizes of k, which should be within the partition size of k + 1. Therefore, Ward's method does not produce a sum of squares as small as that of the k-means method. Between the k-means method and Ward's method, Ward's method gives more accurate results compared to the k-means method. The trade-off for this accuracy aspect is that due to its complexity, Ward's method takes more time to be calculated and shows less error, as shown in the Tables 8 and 9 below, produced by the Azizi et al. in 2019.

Table 8. Time of clustering with different methods [15].

Method	K-means	Linkage-ward
Time (s)	0.37	0.52

Table 9. Relative error between cluster members and their centers in different methods [15].

Method	K-means	Linkage-ward
Relative Error	10.3%	8.2%

Therefore, with regard to the essence of the accuracy of wind clustering, Ward's method shows higher precision compared to the other clustering methods. The method is

also easily applied to numerous parameters, such as speed, direction, frequency, and others, to suit the researcher's target objectives. This paper focuses on the best method of wind clustering according to wind speed. Therefore, it was found that to cluster wind speed at a particular location and a period of time, the clustering should be able to segregate a timelapse, such as with wind speed trend clustering. Table 10 below shows the comparison of each method discussed in this research.

Table 10. Comparison table on clustering method.

Clustering Method	Advantage	Disadvantage
Linkage–Ward clustering method	• Higher accuracy • Available in machine learning software	• Requires more computational effort • More complex calculation
k-means approach	• Ease of data insertion • Easier calculation • Adopted by many researchers • Many improvised versions • Available in machine learning software	• Lesser accuracy
Non-parametric hierarchical clustering approach	• Easier calculation	• Rarely used by researcher • Not readily available in machine learning
Trend-based time series data clustering	• Many improvised versions • Adopted in many research studies	• More complex calculation • Not readily available in machine learning
Anderberg hierarchical clustering method	• Easier calculation	• Rarely used by researchers • Not readily available in machine learning

It concluded that in terms of accuracy, readability in machine learning software, and larger datasets, the most suitable method to cluster the wind trend nationally is the Linkage–Ward clustering method. The selection of the Linkage–Ward clustering method is due to the impact of the result and its accuracy. Although the calculations using Ward's method are more complex than those of the other methods, due to impact of the result the complexity can be ignored. The result of the research aims to create a guideline for researchers, engineers, and wind experts to improve the knowledge and design, especially regarding wind speed trends. The impact of the finding will be on the civil design, wind harvesting, and weather safety sectors.

Funding: This research received no external funding.

Conflicts of Interest: The authors declare no conflict of interest.

References

1. Low, K.C. Application of nowcasting techniques towards strengthening national warning capabilities on hydrometeorological and landslides hazards. In Proceedings of the Public Weather Services Workshop on Warning of Real-Time Hazards by Using Nowcasting Technology, Sydney, Australia, 9–13 October 2006.
2. Satari, S.; Zubairi, Y.; Hussin, A.G.; Hassan, S.F. Some Statistical Characteristic of Malaysian Wind Direction Recorded at Maximum Wind Speed: 1999-2008. *Sains Malays.* **2015**, *44*, 1521–1530. [CrossRef]
3. *Malaysian Standard MS 1553: 2002*; Code of Practice on Wind Loading for Building Structure. Department of Standards Malaysia: Cyberjaya, Malaysia, 2002.
4. Young, I.R.; Zieger, S.; Babanin, A.V. Global Trends in Wind Speed and Wave Height. *Science* **2011**, *332*, 451–455. [CrossRef] [PubMed]
5. Zeng, Z.; Ziegler, A.D.; Searchinger, T.; Yang, L.; Chen, A.; Ju, K.; Piao, S.; Li, L.Z.X.; Ciais, P.; Chen, D.; et al. A reversal in global terrestrial stilling and its implications for wind energy production. *Nat. Clim. Chang.* **2019**, *9*, 979–985. [CrossRef]
6. Chiang, E.P.; Zainal, A.; Aswatha, N.; Seetharamu, K.N. The Potential of Wave and Offshore Wind Energy in Around the Coastline of Malaysia That Face the South China Sea. In Proceedings of the International Symposium on Renewable Energy: Environment Protection & Energy Soilution for Sustainable Development, Kuala Lumpur, Malaysia, 14–17 September 2003.
7. Kok, P.H.; Akhir, M.F.; Tangang, F.T. Thermal frontal zone along the east coast of Peninsular Malaysia. *Cont. Shelf Res.* **2015**, *110*, 1–15. [CrossRef]

8. Stewart, R.H. Physical Oceanography. *Deep Sea Res. Part Oceanogr. Liter. Rev.* **1987**, *34*, 629–645.
9. United States Environmental Protection Agency. Climate Change Indicators: Oceans | US EPA. 2022. Available online: https://www.epa.gov/climate-indicators/oceans (accessed on 14 March 2023).
10. Masseran, N.; Razali, A.; Ibrahim, K.; Zin, W.W. Evaluating the wind speed persistence for several wind stations in Peninsular Malaysia. *Energy* **2012**, *37*, 649–656. [CrossRef]
11. Aboobacker, V.M.; Shanas, P.R.; Veerasingam, S.; Al-Ansari, E.M.; Sadooni, F.N.; Vethamony, P. Long-Term Assessment of Onshore and Offshore Wind Energy Potentials of Qatar. *Energies* **2021**, *14*, 1178. [CrossRef]
12. Tiang, T.L.; Ishak, D. Technical review of wind energy potential as small-scale power generation sources in Penang Island Malaysia. *Renew. Sustain. Energy Rev.* **2012**, *16*, 3034–3042. [CrossRef]
13. Shanmugasundaram, J.; Harikrishna, P.; Gomathinayagam, S.; Lakshmanan, N. Wind, terrain and structural damping characteristics under tropical cyclone conditions. *Eng. Struct.* **1999**, *21*, 1006–1014. [CrossRef]
14. Sokolov, A.; Dmitriev, E.; Maksimovich, E.; Delbarre, H.; Augustin, P.; Gengembre, C.; Fourmentin, M.; Locoge, N. Cluster Analysis of Atmospheric Dynamics and Pollution Transport in a Coastal Area. *Bound. Layer Meteorol.* **2016**, *161*, 237–264. [CrossRef]
15. Azizi, E.; Kharrati-Shishavan, H.; Mohammadi-Ivatloo, B.; Shotorbani, A.M. Wind Speed Clustering Using Linkage-Ward Method: A Case Study of Khaaf, Iran. *GAZI Univ. J. Sci.* **2019**, *32*, 945–954. [CrossRef]
16. Yesilbudak, M. Clustering Analysis of Multidimensional Wind Speed Data Using K-Means Approach. In Proceedings of the 2016 IEEE International Conference on Renewable Energy Research and Applications, ICRERA, Birmingham, UK, 20–23 November 2016; Volume 5, pp. 961–965.
17. Kusiak, A.; Li, W. Short-term prediction of wind power with a clustering approach. *Renew. Energy* **2010**, *35*, 2362–2369. [CrossRef]
18. Clifton, A.; Lundquist, J. Data Clustering Reveals Climate Impacts on Local Wind Phenomena. *J. Appl. Meteorol. Clim.* **2012**, *51*, 1547–1557. [CrossRef]
19. Zhao, Q.; Hautamaki, V.; Fränti, P. Knee Point Detection in BIC for Detecting the Number of Clusters. In *Advanced Concepts for Intelligent Vision Systems: 10th International Conference, ACIVS 2008, Juan-les-Pins, France, October 20–24*; Springer: Berlin/Heidelberg, Germany, 2008; pp. 664–673. [CrossRef]
20. Güldal, V.; Hakan, T. Clustering Analysis in Search of Wind Impacts on Evaporation. *Appl. Ecol. Environ. Res.* **2008**, *6*, 65–73. [CrossRef]
21. Kushwah, V.; Wadhvani, R.; Kushwah, A.K. Trend-based time series data clustering for wind speed forecasting. *Wind. Eng.* **2020**, *45*, 992–1001. [CrossRef]
22. van Vuuren, C.Y.J.; Vermeulen, H.J. Clustering of wind resource data for the South African renewable energy development zones. *J. Energy South. Afr.* **2019**, *30*, 126–143. [CrossRef]
23. Kaufmann, P.; Weber, R. *Transactions on Ecology and the Environment*; WIT Press: Billerica, MA, USA, 1996; Volume 11, ISSN 1743-3541.
24. Gassman, F.; Feller, W.; Kaufmann, P.; Megariti, V.; Kamber, K. Development of a Tool for Air Pollution Management and Emergency Response (MISTRAL Project). *Transact. Ecol. Environ.* **1993**, *1*, 54–60.
25. Angosto, J.M.; Elvira-Rendueles, B.; Bayo, J.; Moreno, J.; Vergara, N.; Moreno-Clavel, J.; Moreno-Grau, S. Wind Classification through Cluster Analysis for the Development of Predictive Statistical Models on Atmospheric Pollution. *Adv. Air Pollut.* **2002**, *11*, 635–644.
26. Dokuz, A.S.; Demolli, H.; Gokcek, M.; Ecemis, A. Year-Ahead Wind Speed Forecasting Using a Clustering-Statistical Hybrid Method. In Proceedings of the International Conference on Innovative Engineering Applications (September), Sivas, Turkey, 20–22 September 2018; pp. 971–975.

Disclaimer/Publisher's Note: The statements, opinions and data contained in all publications are solely those of the individual author(s) and contributor(s) and not of MDPI and/or the editor(s). MDPI and/or the editor(s) disclaim responsibility for any injury to people or property resulting from any ideas, methods, instructions or products referred to in the content.

Article

Predication of Ocean Wave Height for Ocean Wave Energy Conversion System

Yingjie Cui [1], Fei Zhang [1,*] and Zhongxian Chen [1,2]

[1] School of Intelligence Manufacturing, Huanghuai University, Zhumadian 463000, China; cuiyingjie@huanghuai.edu.cn (Y.C.)
[2] Henan Key Laboratory of Smart Lighting, Huanghuai University, Zhumadian 463000, China
* Correspondence: zhangfei@huanghuai.edu.cn

Abstract: Ocean wave height is one of the critical factors to decide the efficiency of the ocean wave energy conversion system. Usually, only when the resonate occurs between the ocean wave height (ocean wave speed in the vertical direction) and ocean wave energy conversion system, can the conversion efficiency from ocean wave energy into electric energy be maximized. Therefore, this paper proposes two predication methods to predict the future ocean wave height in 1.5–2.5 s. Firstly, the data fitting of real ocean wave height is achieved by the polynomial method, which is beneficial to the predication of ocean wave height. Secondly, the models of the moving average (MA) predication method and auto regressive (AR) predication method are presented by the time series analysis process. Lastly, after the predication of ocean wave height by the MA method and AR method, and compared with the data fitting result of real ocean wave height, it can be found that the AR method is more accurate for the predication of ocean wave height. In addition, the predication results also indicated that the error between the predication value and true value in the future 2.5 s is considered acceptable, which provides enough time to optimize the operation process of the ocean wave energy conversion system by a suitable control method.

Keywords: ocean wave height; data fitting; predication method; ocean wave energy conversion

1. Introduction

With the popularization and application of new energy developments, ocean wave energy extraction has become a popular research direction for converting this renewable energy into electric energy [1,2]. However, during the research process of ocean wave energy extraction, the conversion efficiency from ocean wave energy into electric energy is low, which is analyzed and verified by many experts [3–5]. For example, reference [3] has estimated that the power efficiency of ocean wave energy to electric energy is 11.97%, and reference [4] has calculated that the operation efficiency and power translation of the wave energy converter is 23% (without operation control). Moreover, reference [5] has obtained that the natural efficiency of the wave energy conversion system is 1.4%, and the experimental test process was implemented in the Yellow Sea near Lianyungang port.

In order to improve the operation efficiency of the ocean wave energy conversion system, some control methods have been proposed and researched. With the certain structure of the ocean wave energy conversion system, the field weakening (FW) control of the linear generator [6], second order sliding mode control [7–9], model-predictive control (MPC) [10,11], sliding mode control [12] and other control methods have been investigated one after another. Actually, the real-time motion situation of ocean waves is one crucial factor to improve the operation efficiency of the ocean wave energy conversion system, except for the optimized control of the ocean wave energy conversion system. The reason for this statement is that only the resonate occurs between the ocean wave and ocean wave energy conversion system, and the conversion efficiency from ocean wave energy

into electric energy is maximum [13]. Therefore, real-time monitoring or predicting of the motion situation of the ocean wave is beneficial to the optimization control and efficiency improvement of the ocean wave energy conversion system.

At present, the focus of research on ocean wave height predication is mostly based on empirical equations and numerical simulation. In 2021 and 2022, Jrges et al. and Lou et al. proposed a long short-term memory (LSTM) neural network model to forecast ocean wave height [14,15]. The LSTM neural network model is suitable for high and low ocean wave height predication, and its predication accuracy in the short-term period and long-term period of ocean waves was improved by 7.4–11.7% and 8.8–9.1%, respectively. Moreover, some other kinds of ocean wave height predication with the LSTM neural network have been proposed in recent years. Fan et al. proposed a simulating waves nearshore LSTM model to make a single-point predication of ocean wave height, and its performance was outperformed by the standard SWAN model with an over 65% improvement in accuracy [16]. For the higher lead times of ocean wave height, the recurrent neural networks (RNN) based on the correlation coefficient and Root Mean Square Error of the LSTM model perform better than the persistence model [17]. Moreover, ocean wave height predication is beneficial for the wave power density estimate; for example, a modified gamma spectrum method is presented and revised to estimate the wave energy density in a desired geographic area [18], and a series of wave energy estimation methods with the linear wave theory are achieved in the Gulf of Mexico [19].

However, the predication of ocean wave height using empirical equations and numerical simulation is complicated and needs a long calculation time. If the predication time is too long, it will be detrimental to implement the optimal operation control and efficiency improvement of the ocean wave energy conversion system. For the predication of ocean wave height quickly and effectively, this paper proposes two predication methods (the MA and AR predication methods) to predict the future ocean wave height. After the modeling and simulation calculation, it found that both the two predication methods are time-saving, and the AR predication method is more accurate than the MA predication method. It can also accurately predict the future ocean wave height in the next 2.5 s.

This paper was arranged as follows: Section 2 describes the data fitting of real ocean wave height by the polynomial method, which is beneficial to the predication of ocean wave height. Section 3 presents the predication models of the MA method and AR method by the time series analysis processes. Section 4 compares the predication results with the data fitting result of real ocean wave height, and analyzes the predication accuracy of the MA method and AR method. Finally, the discussions and conclusions are drawn in Sections 5 and 6.

2. Data Fitting of Real Ocean Wave Height

Actually, the motion process of real ocean waves is irregular [20]. Figure 1 shows the real ocean waves near the ocean wave energy conversion system. From Figure 1, it can be seen that the real ocean waves are irregular, including the period and height of the ocean waves. Without predicting or measuring such kinds of ocean wave height, it is difficult to maximize the conversion efficiency of ocean wave energy into electrical energy. If the ocean wave height is measured by instruments, many difficulties will occur, such as the installation position of instruments, high-cost investment and so on. Therefore, this paper adopts the time series analysis method to predict the future ocean wave height. In addition, the operational principle of the ocean wave energy conversion system in Figure 1 can be found in the reference [5].

If research is conducted on the height predication of real ocean waves directly, then the actual optimization control of the ocean wave energy conversion system will become complex and even difficult to achieve. Thus, the method of data fitting is introduced in this paper, which aims to find the physical and mathematical significance of ocean wave motion, and provide less but effective data information for the predication of ocean wave height.

Figure 1. The real ocean waves near the ocean wave energy conversion system.

In some papers or books, data fitting is also called curve fitting [21]. Through the method of data fitting, the date series (data in chronological order) can be described by smooth curves or mathematical expressions. Usually, there are five mathematical methods to achieve data fitting, such as the least square method, stepwise regression method, polynomial method, logarithmic method and gamma adjustment method. This paper adopts the polynomial method to process ocean wave height variation with time. The representation of the polynomial function can be written as

$$f(x,w) = w_0 + w_1 x + w_2 x^2 + w_3 x^3 + \cdots + w_m x^m = \sum_{i=0}^{m} w_i x^i \qquad (1)$$

where x is the input variables, $f(x,w)$ is the output variables, w is the coefficient and m is the order of polynomial function.

Using the polynomial method, Figure 2 shows the example of data fitting of real ocean wave height, where the time series of ocean wave height is collected in the Yellow Sea near Lianyungang port, on 1 November 2017. From Figure 2, it can be concluded that the real ocean wave height is nonlinear and irregular, which makes it difficult to predict the future ocean wave height, and the subsequent operation control of the ocean wave energy conversion system will become complex and even difficult. However, after the data fitting of real ocean wave height, the variation of ocean wave height is smooth (the small area oscillations are eliminated), which is beneficial to the predication of ocean wave height, and the conversion of ocean wave energy into electric energy will be improved by a suitable operation control method.

Moreover, for the reason of the additional mass and damping coefficient of the ocean wave energy conversion system (device) in the ocean waves, the local fluctuation of ocean waves will not affect the motion of the ocean wave energy conversion system significantly. Therefore, the data fitting of real ocean wave height is reasonable.

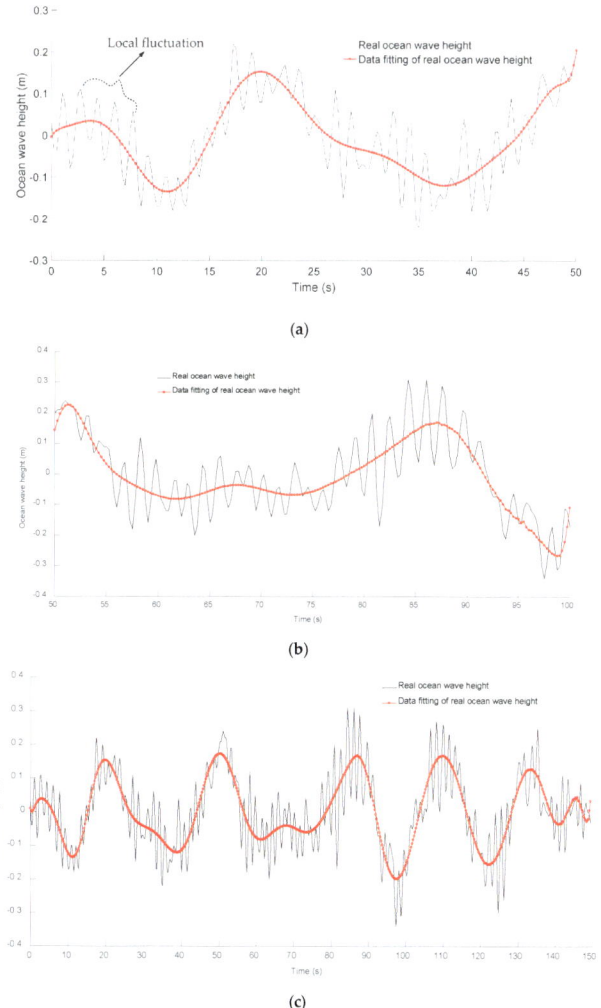

Figure 2. The data fitting of real ocean wave height. (**a**) The time series is 0–50 s. (**b**) The time series is 50–100 s. (**c**) The time series is 0–150 s.

3. Predication Model of Ocean Wave Weight

After the data fitting of real ocean wave height, the future ocean wave height in the next few seconds can be predicted. This paper creates two models to predict the future ocean wave height.

3.1. Moving Average (MA) Model

For the time series x_t, its qth-order MA model can be written as [22]

$$x_t = \mu + \varepsilon_t - b_1 \varepsilon_{t-1} - b_2 \varepsilon_{t-2} - \cdots - b_q \varepsilon_{t-q} \qquad (2)$$

where $b = (b_1, b_2, \cdots b_q)^T$ is the coefficient of the MA model, μ is the invariant constant and $\{\varepsilon_t\}$ is the white noise sequences with $WN(0, \sigma^2)$.

Considering the relationship between predication step l and order q, the future predicted time series x_{t+l} can be written as

$$\begin{aligned} x_{t+l} &= \mu + \varepsilon_{t+l} - b_1\varepsilon_{t+l-1} - b_2\varepsilon_{t+l-2} - \cdots - b_q\varepsilon_{t+l-q} \\ &= (\varepsilon_{t+l} - b_1\varepsilon_{t+l-1} - \cdots b_{l-1}\varepsilon_{t+1}) + \left(\mu - b_l\varepsilon_t - \cdots b_q\varepsilon_{t+l-q}\right) \quad (l \leq q) \end{aligned} \quad (3)$$

$$\begin{aligned} x_{t+l} &= \mu + \varepsilon_{t+l} - b_1\varepsilon_{t+l-1} - b_2\varepsilon_{t+l-2} - \cdots - b_q\varepsilon_{t+l-q} \\ &= \left(\varepsilon_{t+l} - b_1\varepsilon_{t+l-1} - b_2\varepsilon_{t+l-2} - \cdots b_q\varepsilon_{t+l-q}\right) + \mu \quad (l > q) \end{aligned} \quad (4)$$

At the right of the last equal sign of Formulas (3) and (4), the first part is the predication error, and the second part is the predication value. By comparing Formulas (3) and (4), it can be obtained that the predicted value is practical only when the predication step l is smaller than the order q.

In order to improve the accuracy of the MA model, the double moving average method is adopted to process the results of the MA model further, which is described as follows:

$$M_t^{(1)} = \frac{x_t + x_{t-1} + \cdots + x_{t-N+1}}{N} \quad (5)$$

$$M_t^{(2)} = \frac{M_t^{(1)} + M_{t-1}^{(1)} + \cdots + M_{t-N+1}^{(1)}}{N} \quad (6)$$

$$a_t = 2M_t^{(1)} - M_t^{(2)} \quad (7)$$

$$b_t = \frac{2\left(M_t^{(1)} - M_t^{(2)}\right)}{N-1} \quad (8)$$

$$x_{t+l} = a_t + b_t l \quad (9)$$

3.2. Auto Regressive (AR) Model

The pth-order AR model can be described as [23]

$$x_t = a_0 + a_1 x_{t-1} + a_2 x_{t-2} + \cdots + a_p x_{t-p} + \varepsilon_t \quad (10)$$

where $a = (a_0, a_1, a_2 \cdots, a_p)^T$ is the autoregressive coefficient of the AR model, $\{\varepsilon_t\}$ is the white noise sequences with $WN(0, \sigma^2)$. In the Formula (10), if the autoregressive coefficient $a_0 = 0$, then the pth-order AR model is named as zero mean sequence. Formula (10) indicates that future value with k steps can be predicted by the history time series $x_t = (x_1, x_2 \cdots, x_t)$. However, in order to improve the feasibility and accuracy of the predication result, some data analyses and processing should be achieved. Figure 3 shows the data analysis and processing of the AR model.

Step 1: White noise test. The aim of the white noise test is to check the feasibility of future ocean wave height predication by the AR model. If the time series data (ocean wave height) is white noise, then the modeling of the AR model (the ocean wave height cannot be predicted) stops, or else converts to Step 2.

Step 2: Stable test of time series data. If the time series data (ocean wave height) are unstable, then a stable process, such as the differential processing method, should be implemented on the time series data. After the stable process, the white noise test of Step 1 will check the feasibility of the time series data again. When the time series data are stable, then converts to Step 3.

The method of the stable test of time series data in this paper is the Phillips-Perron test [24].

Step 3: Model order. Normally, the order of the AR model can be determined by the autocorrelation coefficient and partial correlation coefficient of the time series data. However, in order to create an AR model with optimal order and optimal residual series variance, the model order selection should be further considered.

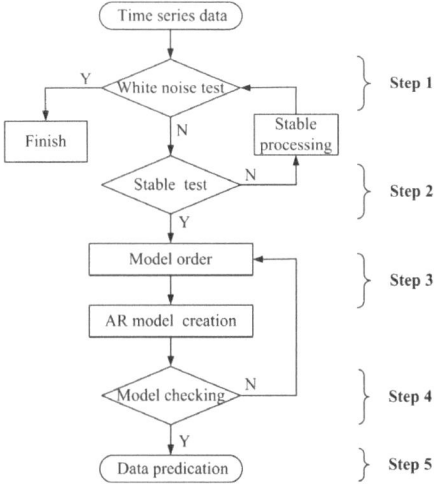

Figure 3. Data analysis and processing of AR model.

This paper adopts the Akaike information criterion (AIC) and the Bayesian information criterion (BIC) to determine the optimal order of the AR model [25].

The AIC criterion can be described as

$$\text{AIC}(p) = \text{Ln}\sigma^2 + \frac{2(p+1)}{N} \tag{11}$$

where $p+1$ is the number of data that will be predicted, σ^2 is the random error variance and N is the number of time series data. When the predicted data $p+1$ increases, the logarithm of random error variance $\text{Ln}\sigma^2$ decreases, but the second term $\frac{2(p+1)}{N}$ in Formula (11) increases. Therefore, under the condition of $p+1$ increases, the value of $\text{AIC}(p)$ can be minimum. After the process of data analysis and comparison, the minimum value of $\text{AIC}(p)$ means the optimal order of the AR model.

The analytic expression of BIC criterion is

$$\text{BIC}(p) = \text{Ln}\sigma^2 + \frac{\text{Ln}2(p+1)}{N} \tag{12}$$

Compared with Formula (11) of the AIC criterion, the last item of Formula (12) of the BIC criterion is Ln2. When the number of time series data N is larger, then Ln2 $>>$ 2; thus, the really suitable orders of the BIC criterion are lower than the orders of the AIC criterion.

Actually, the purpose of the AIC criterion and BIC criterion is to compare the balance point between the data fitting residuals and the number of data, which can provide some choices for us to concentrate on the data fitting error or model complexity.

Normally, the AR model creation is based on the minimum order of the AIC criterion and BIC criterion.

Step 4: Model checking. After AR model creation, the parameter estimation of the AR model and the test of white noise residuals should be established. Firstly, the parameter estimation of the AR model is to estimate the parameters $a = (a_0, a_1, a_2 \cdots, a_p)^T$ of Formula (10) and the random error variance σ^2. Secondly, the test of white noise residuals is to check the feasibility of time series data, which is based on the model order of Step 3.

Step 5: In this step, the future time series data are predicted, and the error between the predication value and real value is compared, which verify the feasibility and accuracy of the AR model in the predication of future ocean wave height.

3.3. An Example of Ocean Wave Height Predication by AR Model

According to the steps of data analysis and processing of the AR model (see Figure 3), an example of ocean wave height predication is investigated, as shown in Figure 4. In Figure 4d, the historical time series (data of ocean wave height) is 0–46.25 s.

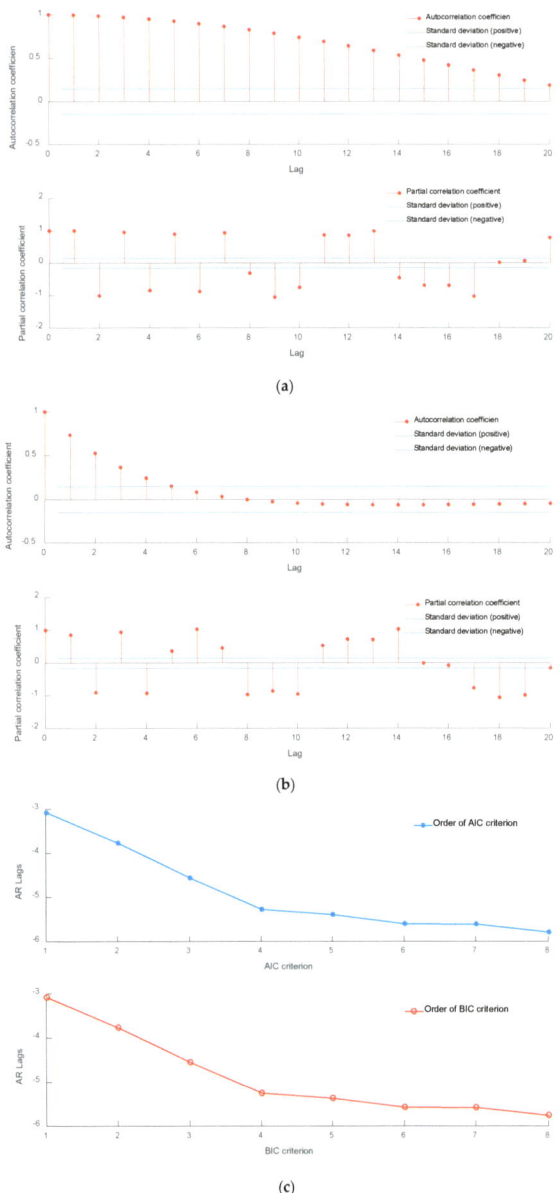

Figure 4. *Cont.*

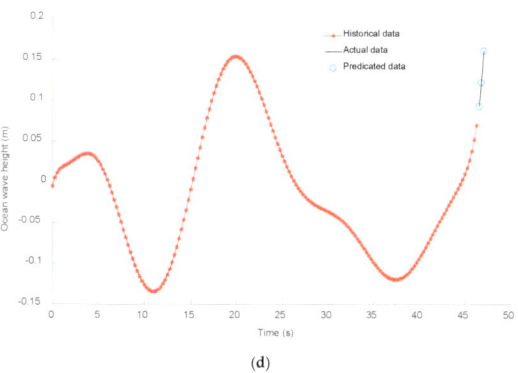

(d)

Figure 4. An example of ocean wave height predication by AR model. (**a**) Autocorrelation coefficient and partial correlation coefficient (before the data are stable). (**b**) Autocorrelation coefficient and partial correlation coefficient (after the data are stable). (**c**) AIC criterion and BIC criterion. (**d**) Ocean wave height predication (three future data are predicated).

As the historical time series of ocean wave height is not the white noise, then the stable test of historical time series is implemented, and the autocorrelation coefficient and partial correlation coefficient are calculated before and after time series stable processing. Compared with Figure 4a, the decay rate of the autocorrelation coefficient of Figure 4b is higher, which suggests that after the time series stable processing, the autocorrelation coefficient of the historical time series of ocean wave height is truncation. However, after the time series stable processing, the partial correlation coefficient is not within the range of two times standard deviation. Therefore, in order to determine the optimal order of the AR model, the AIC criterion and BIC criterion of historical time series are achieved, as shown in Figure 4c. From Figure 4c, it can be concluded that the optimal order range of the AR model is about 4–8, as the variation rate of the Lags value is small in this order range.

After the optimal order of the AR model is determined by the AIC criterion and BIC criterion, an AR model is created, and the predication result of the three future data of ocean wave height is described in Figure 4d. The comparison between the actual data and predicated data means that the created AR model is correct.

4. Predication Results and Analysis

Based on the data fitting of real ocean wave height in Section 2 and the predication model in Section 3, this section compares the predication results of the MA model and AR model and analyzes their predication accuracy.

With the predication cycle as 1.5 s (three predication data in one cycle), Figure 5 shows the predication results of the MA model and AR model, where the continue predication time is 46–140 s. From Figure 5a, it can be seen that the predication accuracy of the MA model is lower in the areas of wave crest (point a1, b1, c1 and d1) and wave trough (x1 and y1). However, compared with the predication results of the MA model, the predication results of the AR model are higher and smoother, which are shown in Figure 5b.

The predication cycle of Figure 5 is 1.5 s, and this short time will not benefit the optimization control technology implementation of the ocean wave energy conversion system. Therefore, the longer predication time of ocean wave height is investigated in this section, as shown in Figure 6. In Figure 6, the continue predication time is also 46–140 s, the predication cycle is 2.5 s and the predicated data are five in one predication cycle. Figure 6a indicates that as the predication cycle increases, the predication accuracy of the MA model is lower than Figure 5a, especially in the wave crest area (point a2, b2, c2 and d2) and wave trough area (x2, y2 and z2). However, under the same condition of the predication cycle (2.5 s), the predication accuracy of the AR model is higher than the MA model, as shown

in Figure 6b. Although the predication cycle increases, the predication results' comparison between Figures 5b and 6b show that the predication accuracy of the AR model is nearly constant.

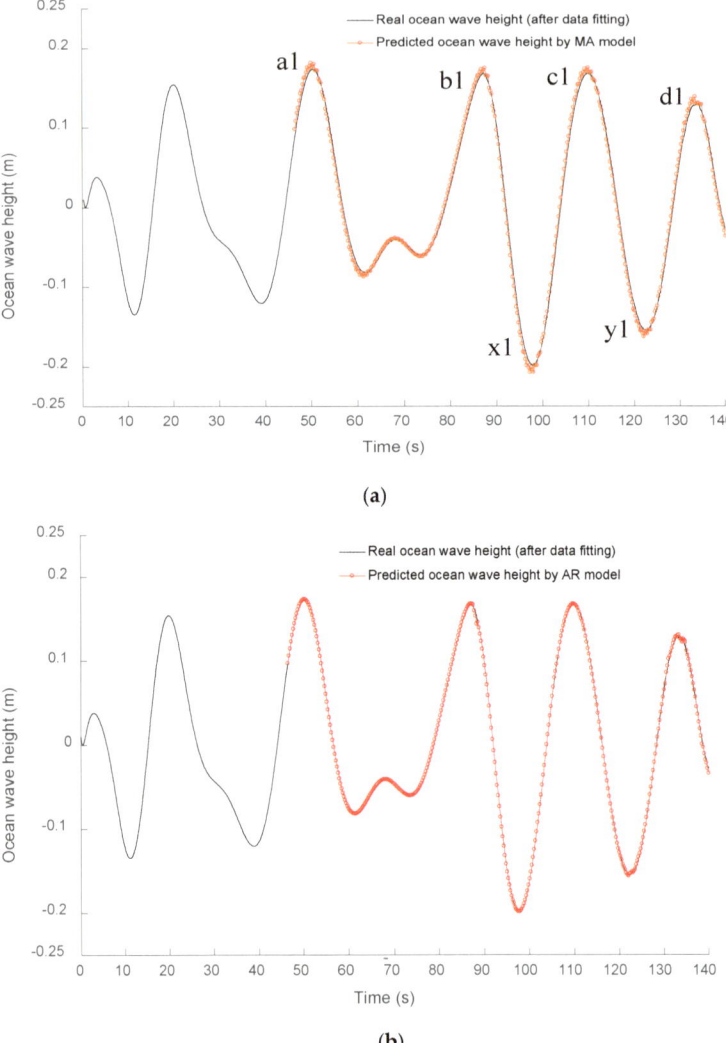

Figure 5. Predication results by MA model and AR model (three future data are a predication cycle). (**a**) Predication result of MA model (predication cycle is 1.5 s). (**b**) Predication result of AR model (predication cycle is 1.5 s).

Figure 7 shows the predication error of the MA model and AR model, where the predication error is the difference value between the real ocean wave height (after data fitting) and the predicted ocean wave height. For the MA model, its predication error is higher in the area of wave crest and wave trough of ocean wave height, which is expressed in Figure 7a. Figure 7a also shows that as the predication cycle increases (from 1.5 s to 2.5 s), the whole predication error is higher.

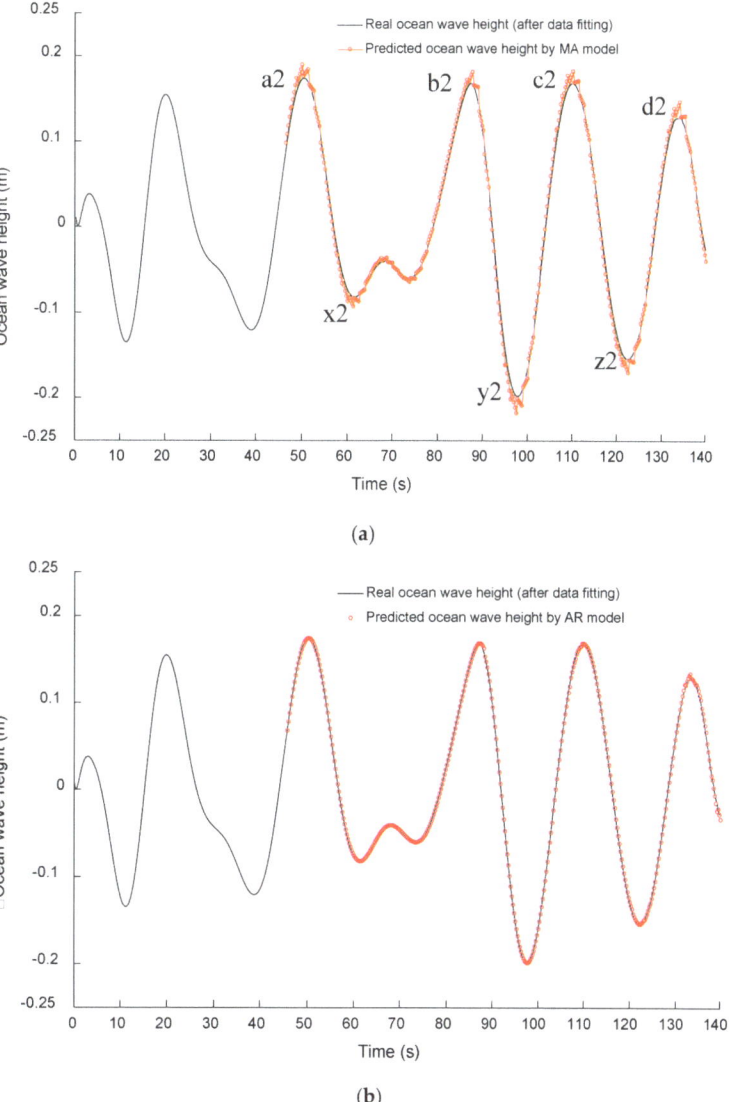

Figure 6. Predication results by MA model and AR model (five future data are a predication cycle). (**a**) Predication result of MA model (predication cycle is 2.5 s). (**b**) Predication result of AR model (predication cycle is 2.5 s).

Figure 7b describes the predication error of the AR model. Compared with the predication error of the MA model, the predication error of the AR model is very small. However, the error distortion occurred at the time of 85–90 s and 130–140 s. The reason for this phenomenon is that in these time intervals, the ocean wave height variation is not smooth, especially in its wave crest area, as shown in Figures 5b and 6b.

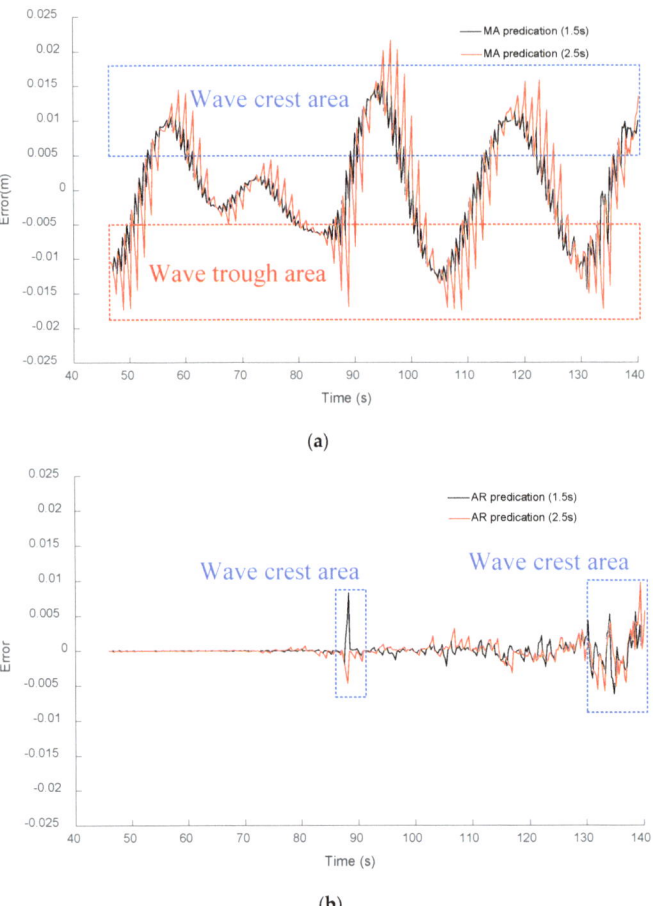

Figure 7. Predication error of MA model and AR model. (**a**) Predication error of MA model. (**b**) Predication error of AR model.

5. Discussion

This paper proposed a data fitting method and two predication methods to increase the predication accuracy of future ocean wave height. Before the conclusion of this paper, some discussions are necessary to be presented.

As shown in Figure 2, the real ocean wave height in one cycle is irregular and displays small-area fluctuation. Therefore, the first discussion concerns the specific impact of this data fitting method on the follow-up optimization control and efficiency of the ocean wave energy conversion system. In order to conduct profound research on this issue, the relationship between ocean waves and wind and the characteristics (period and spectrum et al.) of ocean waves should be investigated. As there are many similarities between the wind and waves (ocean waves), some numerical simulation and analysis methods of wind waves may be beneficial for the predication and research of ocean waves [26–30].

Secondly, although the predication accuracy of the AR model is higher than the MA model, some aberrant error distortions may occasionally occur, as shown in Figure 7b. Thus, a suitable method to avoid or eliminate the aberrant error distortions should be investigated in the next research, which aim is to improve the practicability of the AR model in the predication of ocean wave height.

Lastly, for the structure type of the ocean wave energy conversion system proposed in this paper, the real ocean wave height (historical data and current data) can be measured by the relative motion between the outer buoy and inner buoy, which is beneficial to the real time correction of data fitting, and can also increase the scientific method of the AR model in ocean wave height predication.

In addition, if the predication time is more than 2.5 s, then the white noise sequences and coefficients will be magnified by the calculation process and the predication error will be larger (especially for the ocean waves with both a short wave period and larger wave height), and this phenomenon is not beneficial to the optimization control and efficiency improvement of the ocean wave energy conversion system. However, based on the operational speed of the current controller, 2.5 s is sufficient for the controller to accomplish the optimization control of the ocean wave energy conversion system.

6. Conclusions

In order to improve the efficiency of the ocean wave energy conversion system by a suitable optimization control method, and reduce the complexity and calculation time of ocean wave height predication, this paper proposed two predication methods (the MA and AR predication methods) to predict the future ocean wave height. After the data fitting, modeling and calculation, it found that the AR predication method is more accurate than the MA predication method. It can also accurately predict the future ocean wave height in the next 2.5 s by the AR predication method, which provides enough time to optimize the operation process of the ocean wave energy conversion system by a suitable optimization control method.

Author Contributions: Conceptualization, Z.C. and Y.C.; methodology, F.Z.; writing—original draft preparation, Z.C. and Y.C.; writing—review and editing, F.Z. All authors have read and agreed to the published version of the manuscript.

Funding: This work was financially supported by the Scientific and Technological Project in Henan Province under Grant No. 222102240037 and No. 212102210515.

Institutional Review Board Statement: Not applicable.

Informed Consent Statement: Not applicable.

Data Availability Statement: Some or all data, models generated, or used during the study are available in a repository or online.

Conflicts of Interest: The authors declare no conflict of interest.

References

1. Rizal, A.M.; Ningsih, N.S. Description and variation of ocean wave energy in Indonesian seas and adjacent waters. *Ocean Eng.* **2022**, *251*, 111086. [CrossRef]
2. Khan, M.; Khan, H.A.; Aziz, M. Harvesting energy from ocean: Technologies and perspectives. *Energies* **2022**, *15*, 3456. [CrossRef]
3. Rahman, Y.A.; Setiyawan. The potential of conversion of sea wave energy to electric energy: The performance of central sulawesi west sea using oscillating water column technology. In Proceedings of the Conference Series: Earth and Environmental Science, Bangka Belitung, Indonesia, 29–30 September 2023; Volume 926, p. 012073. [CrossRef]
4. Zou, S.; Zhou, X.; Khan, I.; Weaver, W.W.; Rahman, S. Optimization of the electricity generation of a wave energy converter using deep reinforcement learning. *Ocean Eng.* **2022**, *244*, 110363. [CrossRef]
5. Chen, Z.; Yu, H.; Liu, C.; Hong, L. Design, construction and ocean testing of wave energy conversion system with permanent magnet tubular linear generator. *Trans. Tianjin Univ.* **2016**, *1*, 72–76. [CrossRef]
6. Saenz-Aguirre, A.; Ulazia, A.; Ibarra-Berastegui, G.; Saenz, J. Extension and improvement of synchronous linear generator based point absorber operation in high wave excitation scenarios. *Ocean Eng.* **2021**, *239*, 109844. [CrossRef]
7. Mosquera, F.; Evangelista, C.; Puleston, P.; Ringwood, J.V. Optimal wave energy extraction for oscillating water columns using second order sliding mode control. *IET Renew. Power Gener.* **2020**, *14*, 1512–1519. [CrossRef]
8. Faedo, N.; Mosquera, F.D.; Evangelista, C.A.; Ringwood, J.V.; Puleston, P.F. Preliminary experimental assessment of second-order sliding mode control for wave energy conversion systems. In Proceedings of the Australian & New Zealand Control Conference (ANZCC), Gold Coast, Australia, 24–25 November 2022; pp. 63–68.

9. Gaebele, D.T.; Magaña, M.E.; Brekken, T.K.A.; Henriques, J.C.C.; Carrelhas, A.A.D.; Gato, L.M.C. Second order sliding mode control of oscillating water column wave energy converters for power improvement. *IEEE Trans. Sustain. Energy* **2021**, *12*, 1151–1160. [CrossRef]
10. Zhong, Q.; Yeung, R.W. Model-Predictive control strategy for an array of wave-energy converters. *J. Mar. Sci. Appl.* **2019**, *18*, 26–37. [CrossRef]
11. Ling, B.A.; Bosma, B.; Brekken, T.K.A. Experimental validation of model predictive control applied to the azura wave energy converter. *IEEE Trans. Sustain. Energy* **2020**, *11*, 2284–2293. [CrossRef]
12. Elgammal, A.; Boodoo, C. Optimal sliding mode control of permanent magnet direct drive linear generator for grid-connected wave energy conversion. *Eur. J. Eng. Res. Sci.* **2021**, *6*, 50–57. [CrossRef]
13. Li, Q.; Li, X.; Mi, J.; Jiang, B.; Chen, S.; Zuo, L. A tunable wave energy converter using variable inertia flywheel. *IEEE Trans. Sustain. Energy* **2021**, *12*, 1265–1274. [CrossRef]
14. Jrges, C.; Berkenbrink, C.; Stumpe, B. Predication and reconstruction of ocean wave heights based on bathymetric data using LSTM neural networks. *Ocean Eng.* **2021**, *232*, 109046. [CrossRef]
15. Lou, R.; Wang, W.; Li, X.; Zheng, Y.; Lv, Z. Predication of ocean wave height suitable for ship autopilot. *IEEE Trans. Intell. Transp. Syst.* **2022**, *23*, 25557–25566. [CrossRef]
16. Fan, S.; Xiao, N.; Dong, S. A novel model to predict significant wave height based on long short-term memory network. *Ocean Eng.* **2020**, *205*, 107298. [CrossRef]
17. Pushpam, P.M.M.; Enigo, V.S.F. Forecasting significant wave height using RNN-LSTM models. In Proceedings of the International Conference on Intelligent Computing and Control Systems (ICICCS), Madurai, India, 13–15 May 2020; pp. 1141–1146.
18. Pastor, J.; Liu, Y. Wave climate resource analysis based on a revised gamma spectrum for wave energy conversion technology. *Sustainability* **2016**, *8*, 1321. [CrossRef]
19. Guiberteau, K.L.; Liu, Y.; Lee, J.; Kozman, T.A. Investigation of developing wave energy technology in the Gulf of Mexico. *Distrib. Gener. Altern. Energy J.* **2012**, *27*, 36–52.
20. Kim, S.J.; Kim, M.H.; Koo, W. Nonlinear hydrodynamics of freely floating symmetric bodies in waves by three-dimensional fully nonlinear potential-flow numerical wave tank. *Appl. Ocean Res.* **2021**, *113*, 102727. [CrossRef]
21. Han, B.; Xie, H.; Shan, Y.; Liu, R.; Cao, S. Characteristic curve fitting method of wind speed and wind turbine output based on abnormal data cleaning. *J. Phys. Conf. Ser.* **2022**, *2185*, 012085. [CrossRef]
22. Pemathilake, R.G.H.; Karunathilake, S.P.; Shamal, J.L.A.J.; Upeksh, G. Sales forecasting based on autoregressive integrated moving average and recurrent neural network hybrid model. In Proceedings of the 14th International Conference on Natural Computation, Fuzzy Systems and Knowledge Discovery (ICNC-FSKD), Huangshan, China, 28–30 July 2018; pp. 27–33.
23. Ozawa, M.; Ohtsuki, T.; Jiang, W.; Takatori, Y. Interference alignment for time-varying channel with channel and weight predications based on auto regressive model. In Proceedings of the IEEE Global Communications Conference (GLOBECOM), San Diego, CA, USA, 6–10 December 2015; p. 15820411.
24. Sharma, R.R.; Kumar, M.; Maheshwari, S.; Ray, D.K.P. Evdhm-arima-based time series forecasting model and its application for COVID-19 cases. *IEEE Trans. Instrum. Meas.* **2020**, *70*, 1–10. [CrossRef]
25. Noura, D.; Melita, H. Akaike and bayesian information criteria for hidden markov models. *IEEE Signal Process. Lett.* **2019**, *26*, 302–306.
26. Komori, S.; Kurose, R.; Iwano, K.; Suzuki, N. Direct numerical simulation of wind-driven turbulence and scalar transfer at sheared gas–liquid interfaces. *J. Turbul.* **2010**, *11*, N32. [CrossRef]
27. Sun, H.; Wang, J.; Lin, H.; He, G.; Zhang, Z.; Gao, B.; Jiao, B. Numerical study on a cylinder vibrator in the hydrodynamics of a wind–wave combined power generation system under different mass ratios. *Energies* **2022**, *15*, 9265. [CrossRef]
28. Lin, M.Y.; Moeng, C.H.; Tsai, W.T.; Sullivan, P.P.; Belcher, S.E. Direct numerical simulation of wind-wave generation processes. *J. Fluid Mech.* **2008**, *616*, 1–30. [CrossRef]
29. Yang, S.; Duan, S.; Fan, L.; Zheng, C.; Li, X.; Li, H.; Xu, J.; Wang, Q.; Feng, M. 10-year wind and wave energy assessment in the north Indian ocean. *Energies* **2019**, *12*, 3835. [CrossRef]
30. Li, T.; Shen, L. The principal stage in wind-wave generation. *J. Fluid Mech.* **2022**, *934*, A41. [CrossRef]

Disclaimer/Publisher's Note: The statements, opinions and data contained in all publications are solely those of the individual author(s) and contributor(s) and not of MDPI and/or the editor(s). MDPI and/or the editor(s) disclaim responsibility for any injury to people or property resulting from any ideas, methods, instructions or products referred to in the content.

Article

Numerical Study on Auxiliary Propulsion Performance of Foldable Three-Element Wingsail Utilizing Wind Energy

Yongxu Jiang [1], Chenze Cao [1], Ting Cui [2], Hao Yang [2,*] and Zhengjun Tian [1]

[1] Innovation and R&D Department, China Merchants Industry Technology (Shanghai) Co., Ltd., Shanghai 200137, China; jiangyongxu@cmhk.com (Y.J.); caochenze@cmhk.com (C.C.); tianzhengjun1@cmhk.com (Z.T.)
[2] School of Mechatronics Engineering, Harbin Institute of Technology, Harbin 150001, China; cuiting@hit.edu.cn
* Correspondence: yanghao7196@163.com

Abstract: Sail-assisted propulsion is an important energy-saving technology in the shipping industry, and the development of foldable wingsails has recently become a hot topic. This type of sail is usually composed of multiple elements, and its performance at different folding configurations is very sensitive to changes in incoming airflow, which result in practical operational challenges. Therefore, original and optimized three-element wingsails (bare and concave) are modeled and simulated using the unsteady RANS method with the k-ω SST turbulence model. Next, certain key design and structural parameters (such as angle of attack, apparent wind angle, and camber) are employed to characterize the auxiliary propulsion performance, and the differences are explained in combination with the flow field details. The results show that, in the unfolded state, the aerodynamic performance of the concave wingsail is better than that of the bare wingsail, exhibiting higher lift coefficients, lower drag coefficients, and a more stable surface flow. In the fully folded state, wherein both the nose and flap are rotated, the thrust performance of the concave wingsail remains superior. Specifically, at an angle of attack of 8 degrees, the thrust coefficient of the concave wingsail is approximately 23.5% higher than that of the bare wingsail, indicating improved wind energy utilization. The research results are of great significance for engineering applications and subsequent optimization design.

Keywords: wingsail; apparent wind angle; camber; thrust coefficient

Citation: Jiang, Y.; Cao, C.; Cui, T.; Yang, H.; Tian, Z. Numerical Study on Auxiliary Propulsion Performance of Foldable Three-Element Wingsail Utilizing Wind Energy. *Energies* **2024**, *17*, 3833. https://doi.org/10.3390/en17153833

Academic Editor: Davide Astolfi

Received: 31 May 2024
Revised: 22 July 2024
Accepted: 26 July 2024
Published: 3 August 2024

Copyright: © 2024 by the authors. Licensee MDPI, Basel, Switzerland. This article is an open access article distributed under the terms and conditions of the Creative Commons Attribution (CC BY) license (https://creativecommons.org/licenses/by/4.0/).

1. Introduction

In April 2018, the International Maritime Organization (IMO) adopted the initial strategy for Greenhouse Gas (GHG) emissions, establishing targets and measures to reduce GHG emissions from international shipping [1]. In July 2023, during the 80th session of the Marine Environment Protection Committee (MEPC 80), the initial GHG strategy underwent its first revision, aiming to curb GHG emissions from vessels. The revised strategy set new targets, including a 20% emission reduction by 2030, a 70% reduction by 2040 (relative to 2008 levels), and achieving net-zero emissions by 2050 [2]. Within this context, wingsail-assisted propulsion technology received widespread attention in the shipping and shipbuilding industries for its environmentally friendly and sustainable advantages. This technology utilizes renewable wind energy, decreasing fuel consumption without generating additional onboard carbon emissions [3,4]. As sustainability objectives in shipping intensify, further research into innovative wingsail designs may enable greater adoption of this clean propulsion solution.

In recent years, the development of wingsail-assisted propulsion technology has gradually begun to consider the folding property in both commercial applications and academic research. In August 2023, the foldable two-element wingsail "Wing 560" (Figure 1), designed by Oceanbird, received Approval in Principle (AiP) from the classification society DNV [5]. The wingsail has a height of 40 m, a width of 14 m, and a total sail area of

560 square meters. It consists of two elements, the main body and the flap, and can optimize aerodynamic forces by creating camber profiles. As for the performance, one wingsail on an existing RoRo vessel at normal speed can reduce fuel consumption from the main engine by 7–10% on favorable oceangoing routes. This means a saving of 600 tons of diesel per year, which corresponds to approx. 1920 tons of CO_2. In October 2023, a bulker was equipped with four foldable three-element wingsails, "WindWings" (Figure 2), which was designed by BARTech, aiming to use wind power to reduce fuel consumption and CO_2 emissions [6]. Each of the four "WindWings" measures 20 m in width and 37.5 m in height, and the total surface area of the four wings is 3000 square meters. It consists of three elements, the nose, the main body, and the flap, and can also create cambers by rotating the nose and flap. As for the performance, the "WindWings" can save up to 20% fuel, reducing CO_2 emissions by 19.5 tons per day on an average worldwide route. The similarity between the "Wing 560" and "WindWings" is that both can form different folding configurations through camber angles, thereby fully utilizing the auxiliary performance. However, the performance is very sensitive to the incoming airflow, resulting in difficulties in practical operations; therefore, research on such issues is necessary.

Figure 1. A RoRo vessel with "Wing 560".

Figure 2. A bulker with "WindWings".

Currently, two main approaches are used for studying foldable wingsail performance: model tests and numerical simulations. In 2015, Furukawa [7] conducted a model test to measure the effect of gap geometry, angle of attack, and camber on the performance characteristic of a two-element wingsail, which consists of two different symmetrical airfoils (NACA0025, NACA0009). The gap size and pivot point of the rear element were found to have only a weak influence on the lift and drag coefficients, while increasing the camber can effectively increase the lift coefficient. However, too high a camber will lead to a certain increase in drag.

With advances in computer technology, wingsail performance can be studied using CFD simulation tools, offering a low-cost solution in terms of time and equipment compared to model tests [8]. In 2020, Li [9] conducted a comparative numerical study of a two-element wingsail using steady and unsteady Reynolds-Averaged Navier–Stokes (RANS) methods. The analysis included an examination of the aerodynamic performance at different bends, flap rotational axis positions, angles of attack, and flap thicknesses, ultimately revealing

the nonlinear coupling effect between the bends and flap rotational axis positions. In 2023, Li [10] numerically investigated the aerodynamic performance of a two-element wingsail under gradient and uniform winds, discovering that the gradient wind condition could delay instances of stalling.

However, most previous experimental and numerical simulations have been conducted on two-element wingsails, and very little information is available in the open peer-reviewed literature of foldable three-element wingsails. In comparison, the folding configuration of the three-element wingsail is more complex when considering the cambers of sub-wings. Therefore, an original three-element wingsail will be modeled, and the design will be further optimized to further improve the efficiency of wind energy utilization. Firstly, the aerodynamic and thrust performance of wingsails in the unfolded state will be studied, and the effects of AOA and the apparent wind angle on the lift, drag, and thrust coefficients will be evaluated. Then, considering the foldable properties of wingsails, the thrust performance will be evaluated for a single rotation of the nose and flap separately. Finally, both the nose and flap will be rotated to find the target folding configuration that produces the maximum thrust coefficient. Furthermore, aerodynamic and thrust performance results will be comprehensively analyzed by considering corresponding flow detail characteristics.

2. Model and Methods

2.1. Geometry

Referring to the cross-section of the "WindWings" described in the Introduction, the bare three-element wingsail is first modeled, and the geometry is obtained by stretching three NACA0012 airfoil profiles, as shown in Figure 3a. In addition, with reference to the "slotted flap" design of a two-element sail introduced by Blakeley [11], the airfoil profile of the bare three-element wingsail is optimized, resulting in a concave three-element wingsail, as shown in Figure 3b. Although this wingsail also consists of three parts, as a whole, it has the same shape as the single airfoil pattern. It should be noted that the cross-section type of both the bare and concave three-element wingsail is further designed based on the NACA0012 airfoil profile, and a detailed description of the configuration and parameters will be carried out in the following content.

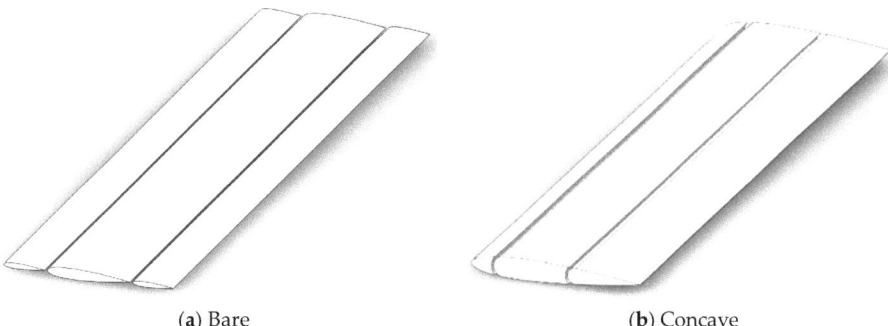

(a) Bare   (b) Concave

Figure 3. Three-dimensional model of the wingsails.

As shown in Figure 4, both the bare and concave three-element wingsails consist of three parts—a nose, main body, and flap—with their specific dimensional parameters detailed in Table 1. To ensure a consistent variable for the comparative study, the total chord length (c) of each wingsail is set to 1 m. The nose, main body, and flap have chord length ratios of 1:2:1 for the bare design and 1:3.7:3.7 for the concave design. In addition, gaps are present between neighboring portions of the wingsails. To mitigate the impact of gap size on the study results, all gap sizes are set to be the same. It should be noted that

both the nose and flap of the bare and concave wingsails can be rotated at an angle to create different cambers, resulting in different folding configurations, which will be described in Section 3.2.

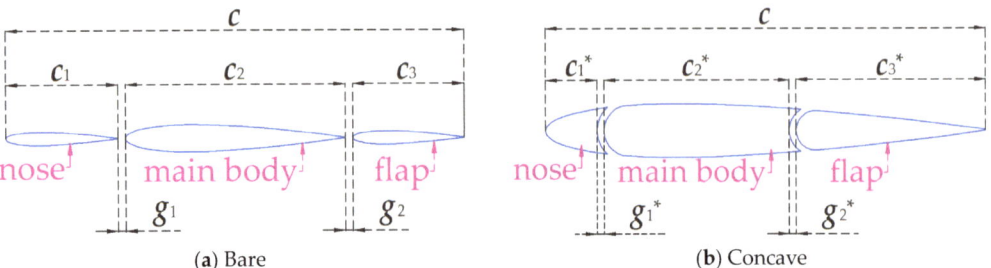

Figure 4. Parameterization of the wingsails.

Table 1. Parameterization of the wingsails.

Parameter	Value
c	1 m
$c_1:c_2:c_3$	1:2:1
$c_1^*:c_2^*:c_3^*$	1:3.7:3.7
g_1, g_1, g_1^*, g_2^*	0.016 m

2.2. Force Analysis of Wingsail

A wingsail is a rigid sail, similar to a wing, that utilizes the Bernoulli effect to generate aerodynamic forces. The airfoil profile allows airflow to pass through, creating a velocity difference, which in turn generates a pressure difference, i.e., lift. This lift, along with the friction component in the direction of sailing, constitutes the propulsive force (as shown in Figure 5). When a vessel is sailing, the wind acting on the surface of the wingsails consists of three components: apparent wind, true wind, and vessel wind. The direction of vessel wind is opposite to the vessel's sailing direction, and the apparent wind is obtained by synthesizing the true wind and vessel wind vectors. The thrust can be determined by synthesizing the lift and drag components in the heading direction, where α is the angle of attack (AOA), β is the apparent wind angle (AWA), V_T is the true wind speed (TWS), V_A is the apparent wind speed (AWS), and V_S is the wind speed caused by vessel sailing.

Focusing on thrust coefficients provides a more directly applicable perspective compared with individual lift and drag values when evaluating wingsail performance for maritime propulsion [12]. Based on the lift and drag coefficients (as shown in Equations (1) and (2)), the thrust coefficient can be calculated by incorporating the apparent wind angle (as shown in Equation (3)). Here, ρ represents the density of air and U denotes AWS. The reference area in the lift and drag coefficients can be replaced by the total chord length of the wingsail in a two-dimensional study [13].

$$C_L = \frac{F_L}{0.5\rho U^2 c} \tag{1}$$

$$C_D = \frac{F_D}{0.5\rho U^2 c} \tag{2}$$

$$C_T = C_L \sin\beta - C_D \cos\beta \tag{3}$$

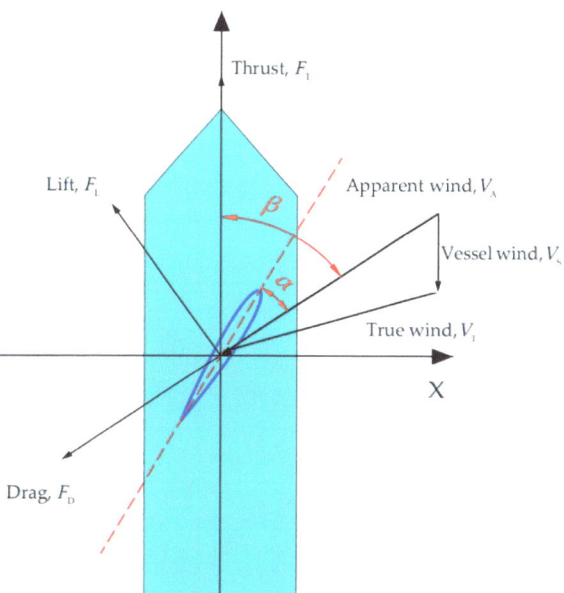

Figure 5. Wind triangle and loads on the wingsail.

2.3. Numerical Setup

2.3.1. Computational Domain

Subsequent numerical simulations are conducted using Star CCM+, a CFD calculation program developed based on the finite volume method (FVM) [14]. In this simulation program, appropriately generating the mesh and configuring the boundary layer settings are critical prerequisites. As depicted in Figure 6, a two-dimensional rectangular fluid computational domain is established with a length of 45 c and width of 30 c. The top and bottom boundaries are symmetric planes aimed at minimizing sidewall effects, while the right side serves as a pressure outlet. Additionally, the sail surface is modeled as a fixed no-slip wall.

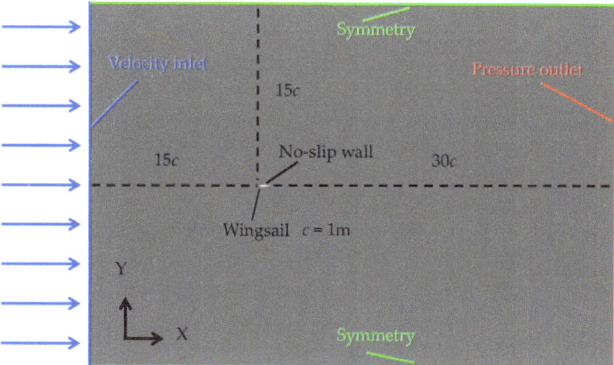

Figure 6. Dimensions of the computational domain and boundary conditions.

Choosing an appropriate turbulence model is equally important when solving the flow field around the wingsail. Hassan [15] used standard, RNG, and realizable $k - \varepsilon$

models and standard and SST $k - \omega$ models to simulate the aerodynamic performance of an NACA0018 airfoil; compared it with experimental results; and found that the SST $k - \omega$ model can achieve more accurate predictions. In STAR-CCM+, the SST $k - \omega$ model is a turbulence model that is used alongside a RANS simulation. This model shares similarities with the $k - \varepsilon$ model, with the main difference being that the SST $k - \omega$ model uses a hybrid approach. If the cells are close to the wall, the $k - \omega$ model will be applied, and if the cells are far away in the free stream, then the $k - \varepsilon$ model will be applied.

2.3.2. Mesh Generation and Convergence Test

In STAR-CCM+, three meshers (Surface Remesher, Trimmed Cell Mesher, and Prism Mesher) are used to generate meshes, and the meshes for the near-sail, wake, and downstream regions are refined separately. The circular and trapezoidal zones near the wingsail allow flow detail features to be effectively captured. In addition, by adjusting various mesh parameters (including the base size, the relative target size, the minimum surface size, the number, and the thickness of prism layers), the final mesh structure is obtained, as shown in Figure 7.

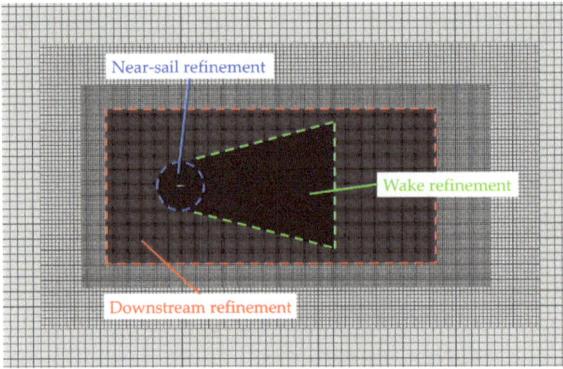

Figure 7. A view of the computational mesh showing refined areas around the geometry.

In the aerodynamic simulation of a wingsail, the meshes of the boundary layer have a significant influence on the results. This layer helps the solver to accurately resolve the flow near the wall, which is vital for determining forces or other flow features, such as separation [16]. Therefore, the prism layer mesher is used to generate the boundary layer. The number of layers is set to 20, the prism layer thickness is set to 0.01 m, and the thickness of the first layer is set to 1.0×10^{-5} m (as shown in Figure 8). As shown in Figure 9, the mean value of Y+ values for both bare and concave three-element wingsail surfaces is less than 1, proving that the boundary layer meshes meet the requirement of simulation accuracy.

In order to accurately capture the time-dependent flow simulation, the Courant–Friedrichs–Lewy (CFL) is introduced (as Equation (4)), where U is the velocity of the fluid, Δt is the time step, and Δx is the characteristic length of the mesh. Referring to the study of Li [9], the mesh size of the leading-edge curve of the flap is taken as the characteristic length of the mesh (i.e., 0.005 m), and the time step used is 0.0002 s. Therefore, the value of CFL is less than 1, which satisfies the requirement of solution accuracy.

$$\text{CFL} = \frac{U \Delta t}{\Delta x} \quad (4)$$

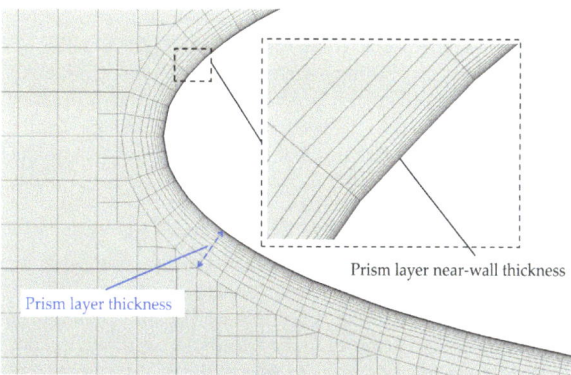

Figure 8. Prism layer close to geometry curves.

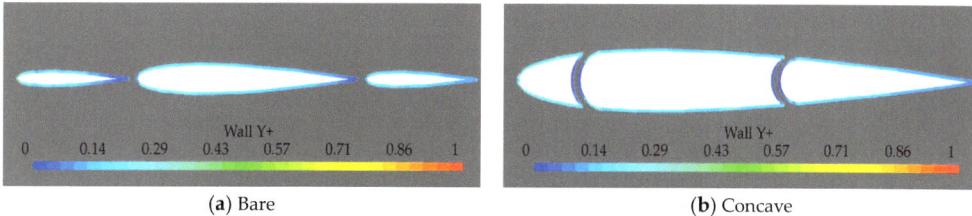

(**a**) Bare (**b**) Concave

Figure 9. Wall Y+ distribution around two types of three-element wingsails.

Furthermore, mesh convergence tests are conducted to assess the sensitivity of the numerical results to the mesh size, with the lift and drag coefficients used as the validation parameters. Three mesh densities (coarse, medium, and fine) are examined, and the total number of meshes and specific results are given in Table 2. The results show that for a bare three-element wingsail, the difference between the results of the coarse mesh and the fine mesh is 7.14–7.44%, while the difference between the medium mesh and the fine mesh is reduced to 0–1.65%. For the concave three-element wingsail, compared with the results of the fine grid, the errors of the coarse and medium mesh are 1.75–5.88% and 0–0.58%, respectively. Overall, the calculation results of the medium grid reached convergence, and for the sake of computational efficiency, this grid size will be used in subsequent simulations.

Table 2. Mesh convergence test (AOA = 2 deg, Re = 700,000).

Model	Case	Mesh Number [-]	C_L	Difference of C_L	C_D	Difference of C_D
bare	coarse	79,775	0.112	7.44%	0.015	7.14%
	medium	135,177	0.119	1.65%	0.014	0
	fine	191,605	0.121	——	0.014	——
concave	coarse	81,015	0.174	1.75%	0.018	5.88%
	medium	137,130	0.172	0.58%	0.017	0
	fine	193,963	0.171	——	0.017	——

2.3.3. Validation

Selecting a validated model for preliminary CFD studies provides confidence in the numerical methods before applying them to the wingsail cases under investigation. Due to the fact that both the bare and concave three-element wingsails are designed based on the

NACA0012 airfoil and the publicly available literature on three-element wingsail model experiments is limited, an NACA0012 airfoil from the wind tunnel test by Sheldahl [17] is selected as the numerical model. The simulation tests are conducted at AOA from 0 to 17 degrees, with a Reynolds number of 700,000. The numerical simulation results in this paper are compared with the experimental results, as shown in Figure 10. The results indicate that within an AOA range of 0 to 10 degrees, the values and trends of the lift coefficient and drag coefficient are in good agreement with the experimental results. However, beyond an AOA of 11 degrees, the numerical simulation results exhibit significant deviations, attributed to the unsteady flow behavior at higher AOAs, particularly beyond the stall. Therefore, the subsequent numerical simulations in this paper are conducted for AOAs ranging from 0 to 10 degrees.

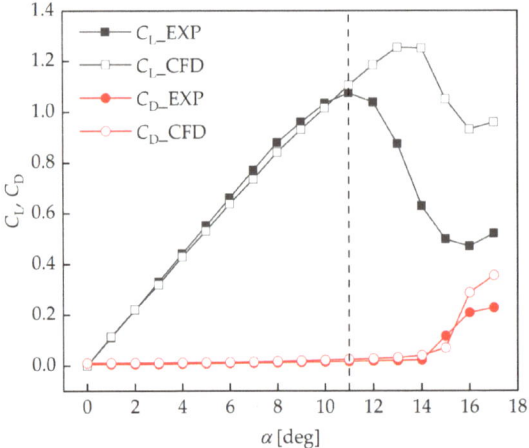

Figure 10. Comparison of lift and drag coefficients between CFD and experiment.

3. Numerical Results

After verifying the feasibility of the numerical approach, applying the validated CFD methodology to the wingsails can reveal quantified aerodynamic differences between the bare and concave design. Firstly, the aerodynamic and auxiliary propulsion performance of unfolded wingsails is studied, and its mechanism is analyzed from the corresponding flow field. Then, considering the foldable properties of the wingsails, the thrust performance is evaluated for a single rotation of the nose and flap separately. Finally, both the nose and flap are rotated to find the target folding configuration that produces the maximum thrust coefficient.

3.1. Wingsails without Camber
3.1.1. Lift and Drag Coefficient

Since the aerodynamic performance of the wingsails is sensitive to AOA, the performance of an unfolded bare and concave wingsail at different AOAs is evaluated first. The lift and drag coefficients of the bare and concave wingsails are investigated in the AOA range of 0 to 10 degrees (as shown in Figure 11). The results show that the lift coefficients of both bare and concave wingsails increase in the range of AOA from 0 to 9 degrees, but a decreasing trend occurs at 10 degrees, with maximum lift coefficient values of 0.47 and 0.63 for bare and concave wingsails, respectively. In addition, the drag coefficients all increase with an increasing AOA. However, the analysis reveals that the difference in lift coefficients between the bare and concave wingsail increases significantly at an AOA greater than 4 degrees, and the drag coefficients similarly differ significantly at an AOA greater than

4 degrees, which may be related to the large change in the flow field, as will be further elaborated on later.

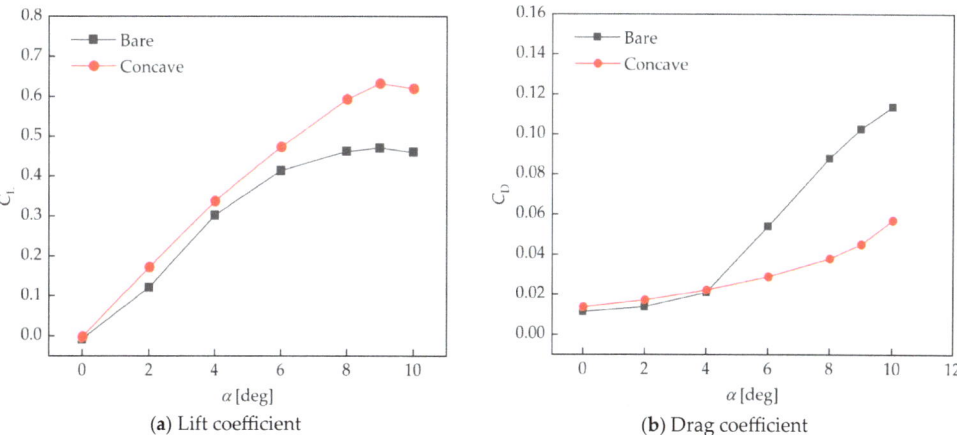

(a) Lift coefficient

(b) Drag coefficient

Figure 11. Force coefficients of the wingsails.

The time history curves of the lift and drag coefficients can help explain the abrupt changes in aerodynamics (as shown in Figures 12 and 13). The results show that the time-course curve of both the lift and drag coefficients of a bare wingsail are straight lines when the AOA is 4 degrees, i.e., the lift and drag coefficients show no fluctuations. However, when the angle of attack is 6, 8, and 10 degrees, the curves of both the lift and drag coefficients show significant fluctuations, which are also due to changes in the flow field pattern around the wingsail. In contrast, the time history curves of the lift and drag coefficients of the concave wingsail do not show any fluctuation at all AOAs, indicating that the flow field pattern around the wingsail is more stable. The aerodynamic performance differences between the bare and concave wingsail will be mechanically explained in the following sections with detailed characteristics of the flow field.

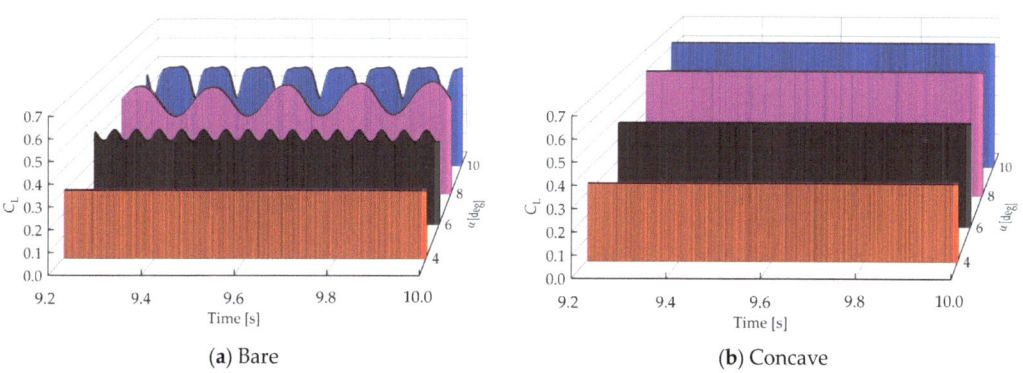

(a) Bare

(b) Concave

Figure 12. Time history curves of lift coefficients of the wingsails.

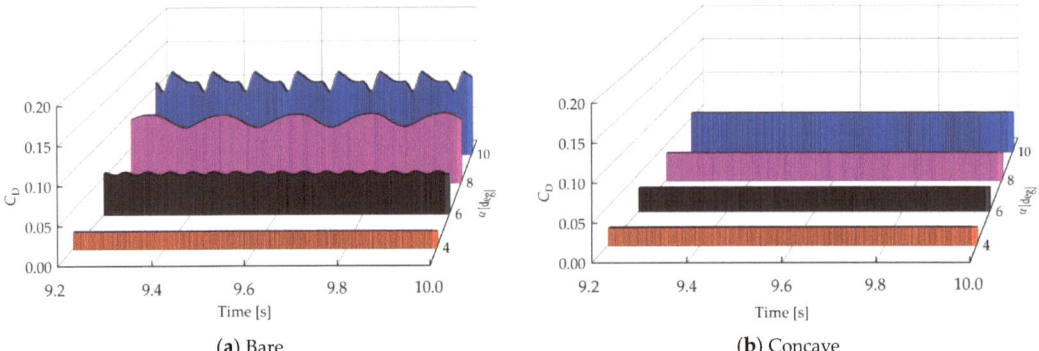

(a) Bare (b) Concave

Figure 13. Time history curves of drag coefficients of the wingsails.

3.1.2. Thrust Coefficient

Focusing on thrust coefficients provides a more directly applicable perspective compared with individual lift and drag values when evaluating wingsail performance for maritime propulsion. As shown in Figure 14, a consistent increasing and then decreasing trend of the thrust coefficients for both bare and concave wingsails occurs at AOAs of 4, 6, 8, and 10 degrees, with peak values occurring around an AWA of 90 degrees. This indicates that wingsails are the most effective in practice when the apparent wind is a crosswind, and the crosswind direction is perpendicular to the vessel heading. In addition, the analysis reveals that the thrust coefficients of the concave wingsail are greater than those of the bare wingsail over most of the AWA range, and this difference increases with an increasing AOA. At an AOA of 10 degrees, the maximum thrust coefficients of the bare and concave wingsails are 0.47 and 0.62, respectively, with a difference of 32%.

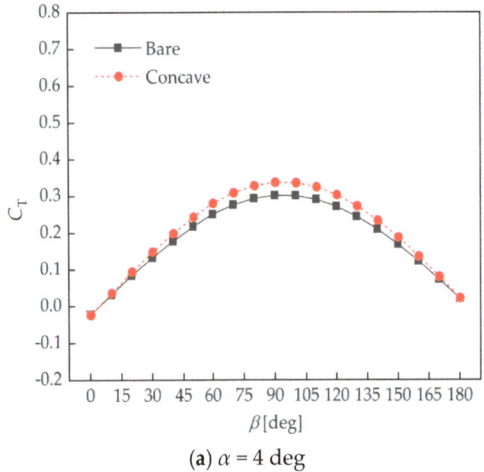

(a) α = 4 deg

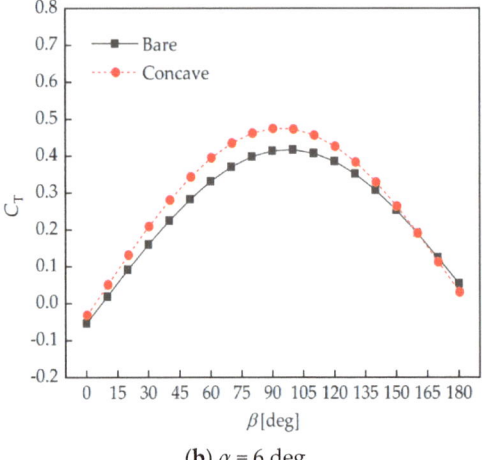

(b) α = 6 deg

Figure 14. Cont.

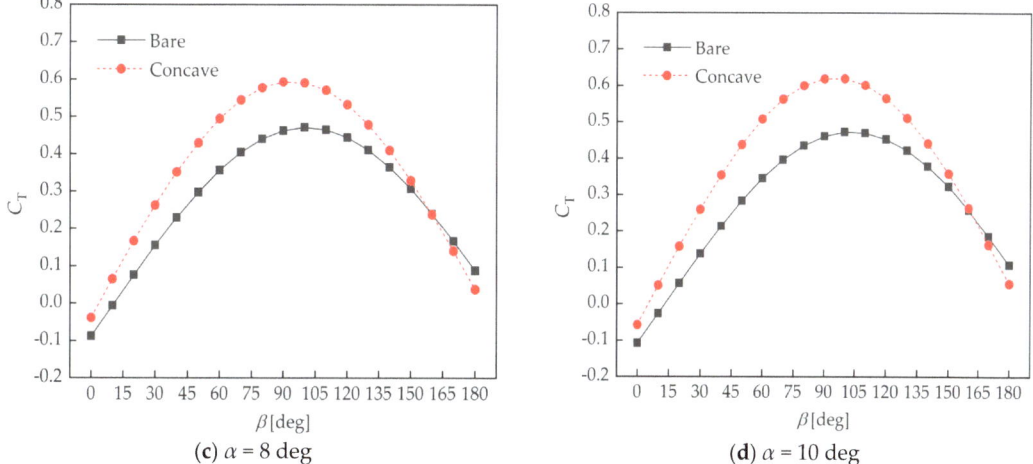

(c) α = 8 deg   (d) α = 10 deg

Figure 14. Thrust coefficients of the wingsails at different AOAs and AWAs.

3.1.3. Flow Pattern

Analyses of the detailed characteristics of the flow field around the wingsails, such as velocity contours and vortex shedding patterns, can provide mechanistic explanations of the previous findings. The flow field patterns around the bare and concave wingsails at AOAs of 4, 6, 8, and 10 degrees are shown in Figure 15. The results indicate that the flow field on the surface of the bare wingsail remains stable at an AOA of 4 degrees. However, when the AOA is increased to 6 degrees, vortex shedding initiates on the suction side of the nose. As the AOA is further increased to 8 and 10 degrees, this vortex shedding pattern becomes more complex and intense, which explains the large fluctuation observed in the force coefficient curves. In contrast, the flow field on the suction side of the concave wingsail remains relatively stable across all AOAs. In addition, at an AOA of 10 degrees, slight flow separation occurs on the suction side of the concave wingsail's flap, resulting in the high-velocity fluid moving away, thereby reducing the pressure difference. This corresponds to the sudden decrease in the lift coefficient (as shown in Figure 8).

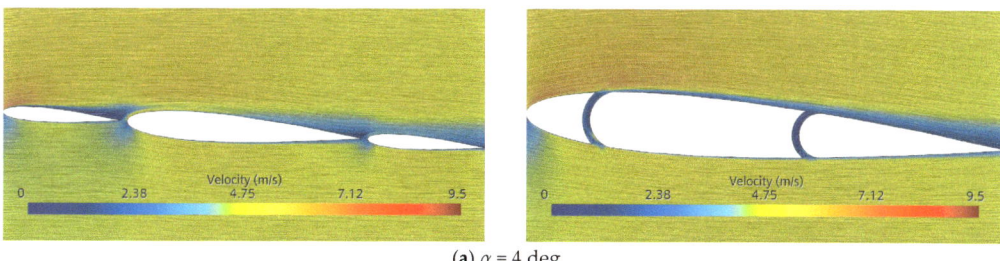

(a) α = 4 deg

Figure 15. *Cont.*

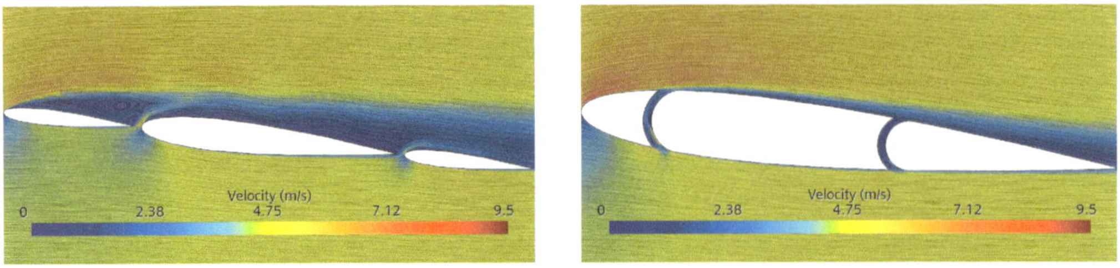

(b) α = 6 deg

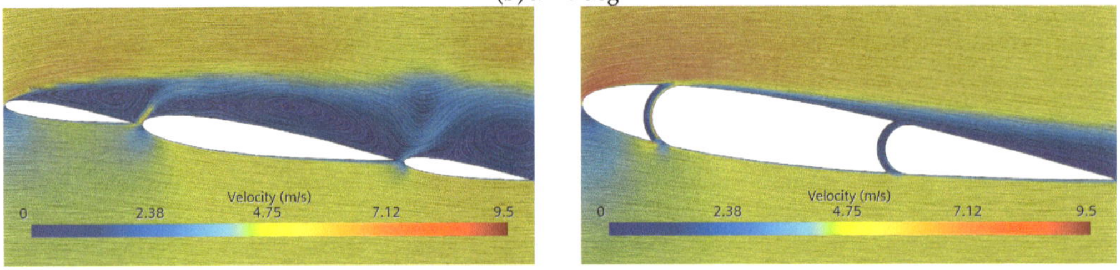

(c) α = 8 deg

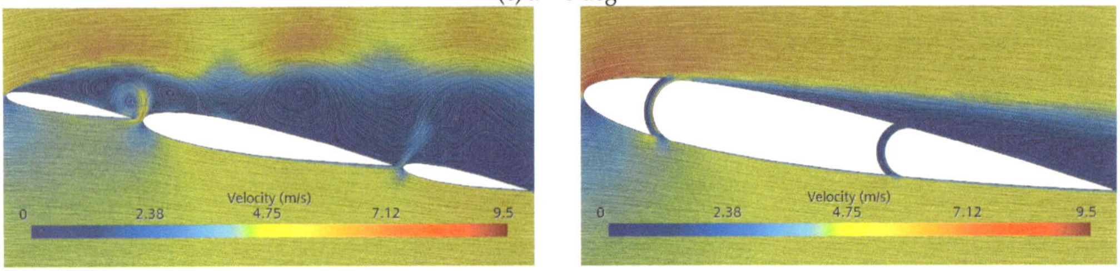

(d) α = 10 deg

Figure 15. Flow pattern around the wingsails at different AOAs.

The velocity contours for the two wingsails are also notably different, as shown in Figure 16. The results indicate that, while the velocity distributions are similar on the pressure side, they significantly differ on the suction side. For the bare wingsail, flow separation occurs at the leading edge of the nose, causing the high-velocity fluid to move away from the surface of the wingsail, resulting in a lower pressure difference and thus a lower lift coefficient. In contrast, for the concave wingsail, flow separation on its suction side occurs at the trailing edge of the nose, leading to the better adsorption of high-speed fluids and a narrower low-speed zone. This difference results in a higher pressure difference between the two sides of the wingsail, thereby leading to a higher lift coefficient, which confirms previous aerodynamic findings.

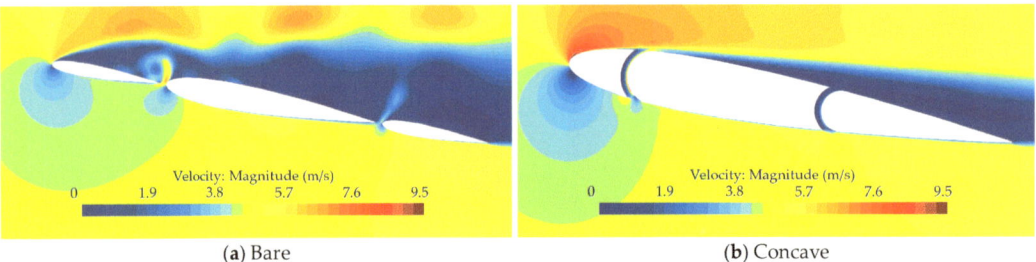

(**a**) Bare (**b**) Concave

Figure 16. Velocity contours of the wingsails (α = 10 deg).

Furthermore, from the flow field of the bare wingsail, it can be seen that the pattern of incoming airflow changes significantly when it is approaching the gaps. In conjunction with the analysis above, it can be seen that it is the larger and wider flow separation zone, which result in some airflow on the pressure side passing through the gaps under negative pressure. In addition to this, this part of airflow will be accelerated by the gaps to rush into the suction side at a higher velocity, which will interfere with the flow separation.

In order to obtain a comprehensive understanding of the aerodynamic performance of both the bare and concave wingsails, we have continued to study the aerodynamic performance at higher AOAs. As shown in Figure 17, it is found that when the AOA is greater than 10 degrees, the vortex shedding patterns on the suction side of the original wingsail (bare wingsail) are similar, while the vorticity intensity increased along with it. Similarly, the steady flow on the suction side of the optimized wingsail (concave wingsail) gradually evolves into a periodic vortex shedding pattern, and the vorticity intensity is also significantly increased. In engineering applications, the high-intensity vortex shedding will cause unstable aerodynamic performance, and the oscillating lift and drag can cause fatigue of the wingsails' structures. In the present study, we focus on the stable and efficient aerodynamic performance of wingsails. Therefore, further research will mainly focus on medium and small attack angles in the following text.

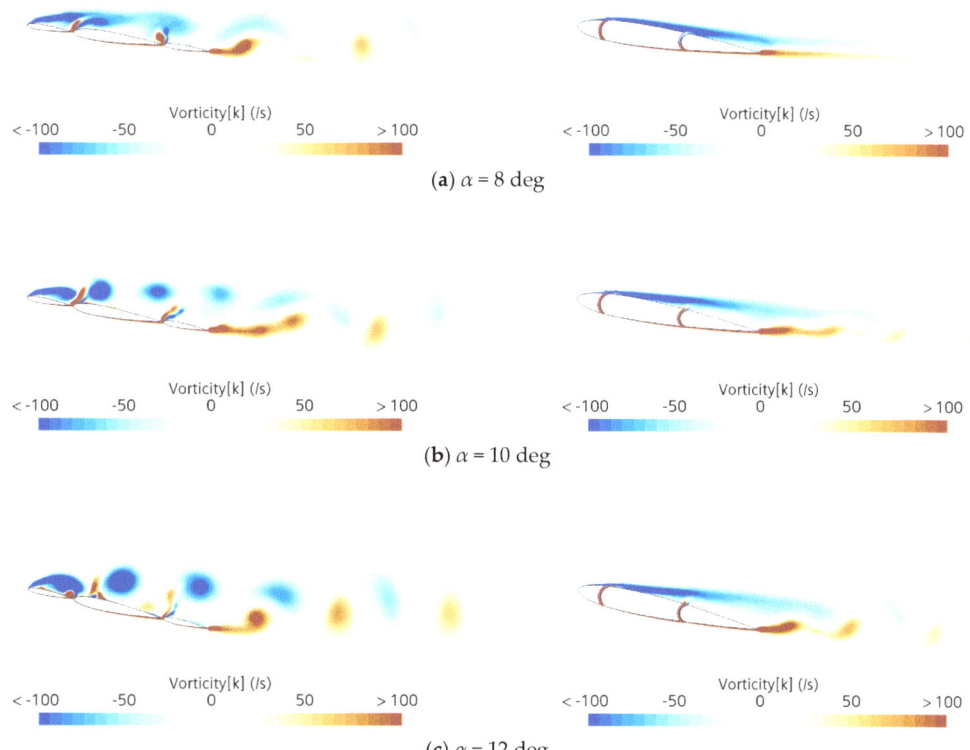

(a) α = 8 deg

(b) α = 10 deg

(c) α = 12 deg

Figure 17. Cont.

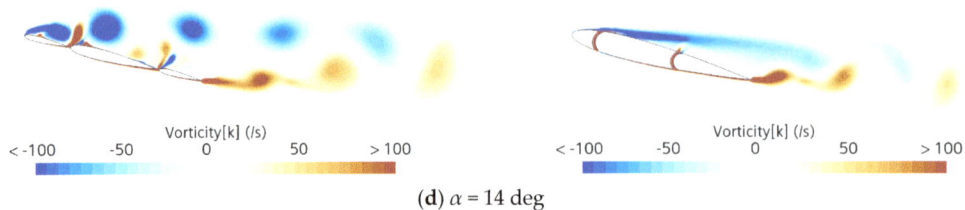

(d) $\alpha = 14$ deg

Figure 17. Vorticity distribution around the wingsails at different AOAs.

3.2. Wingsails with Camber

3.2.1. Definition of Camber

In order to assess the differences between bare and concave wingsails with different folding patterns, a camber is introduced, as depicted in Figure 18. θ_1 and θ_1^* represent the camber angle of the nose, while θ_2 and θ_2^* represent the camber angle of the flap. To simplify this study, the position of the pivot is kept the same for both wingsails, with $x_1 = x_1^* = 0.2$ c, and $x_2 = x_2^* = 0.4$ c. In practical engineering applications, the nose (or flap) and main body are connected by hydraulic telescopic rods, and the two ends of the rods are fixed on the rotating axes of the nose (or flap) and main body, respectively. This mechanism of attachment ensures that the nose and flap can generate cambers to further enhance the thrust performance. The effects of these two components are first evaluated separately in this paper and then collectively in a subsequent study.

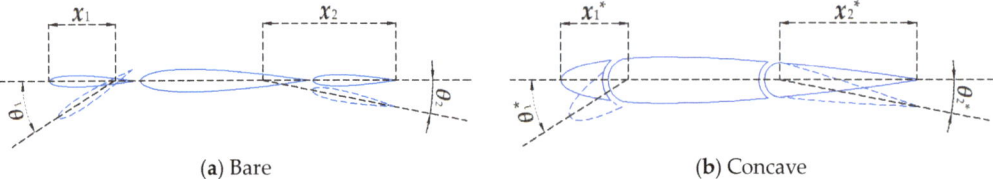

(a) Bare   (b) Concave

Figure 18. Parameterization of the camber and the pivot location of the wingsails.

3.2.2. Individual Evaluation for Nose and Flap Cambers

The thrust coefficients of the two wingsails are evaluated at AOAs of 4, 6, 8, and 10 degrees when rotating the nose and flap separately. Figure 19 shows the thrust coefficient results of bare wingsails and concave wingsails when only the nose rotates. Within the camber range of 0 to 30 degrees, the thrust coefficient of the bare wingsail exhibits a trend of initially increasing and then decreasing. Both the optimal camber angle of the nose and the corresponding thrust coefficient increase with the AOA. Conversely, the thrust coefficient of the concave wingsail shows a monotonically decreasing trend, which indicates that increasing the camber angle of the nose can enhance the thrust performance of the bare wingsail but is detrimental to the concave wingsail. In addition, the trend of the thrust coefficient is generally similar across different AOAs, and an AOA of 8 degrees is selected for a quantitative comparative analysis. The maximum thrust coefficient values for the bare and concave wingsails are achieved at nose camber angles of 10 and 0 degrees, respectively, measuring 0.67 and 0.58, with a difference of 15.5%.

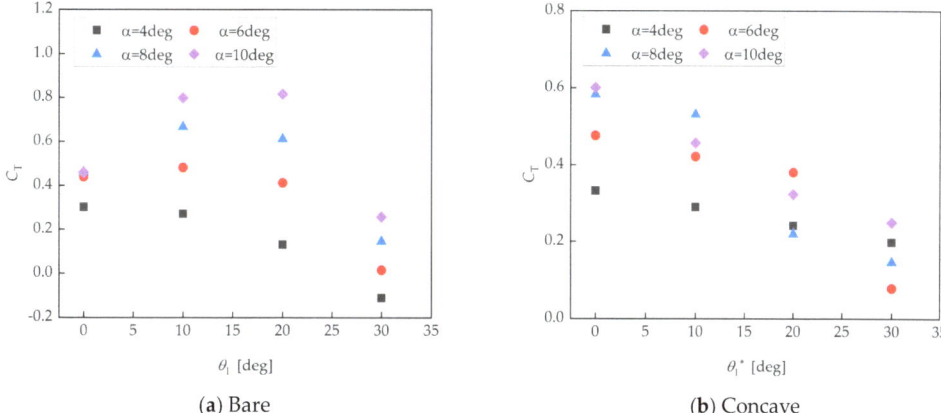

(a) Bare

(b) Concave

Figure 19. Thrust coefficient of the wingsails at different cambers of the nose.

Figure 20 shows the thrust coefficient results of bare and concave wingsails with only the flap rotating. Within the camber range of 0 to 70 degrees, the thrust coefficients of both the bare and concave wingsails exhibit a trend of initially increasing and then decreasing. This suggests that rotating the flap enhances the thrust performance of both the bare and concave wingsails, and this effect is more significant than that of the nose. Similarly, quantitative comparisons are conducted at an AOA of 8 degrees. The analysis shows that the bare and concave wingsails achieve maximum values of 1.47 and 2.06 at a flap camber angle of 50 degrees, respectively, representing a difference of 40.1%. In addition, the thrust coefficient is improved by 19.0% and 53.5%, respectively, compared with the unfolded configuration.

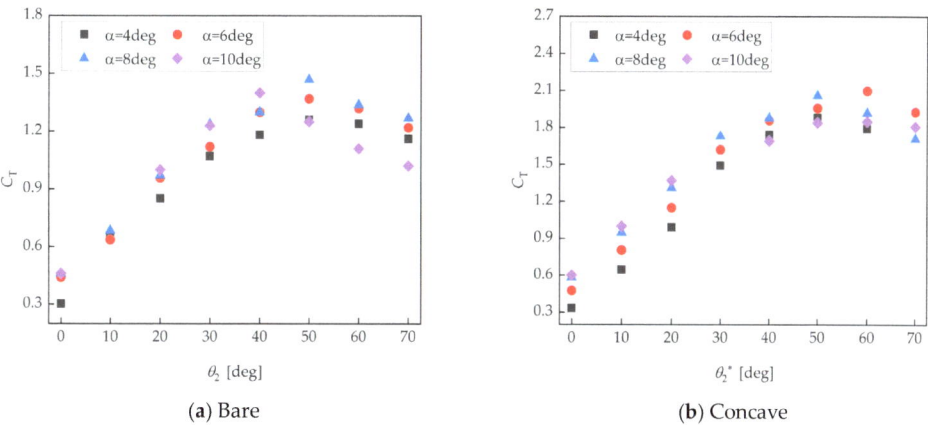

(a) Bare

(b) Concave

Figure 20. Thrust coefficient of the wingsails at different cambers of the flap.

3.2.3. Parallel Evaluation for Nose and Flap Cambers

In order to explore the folding configuration of the bare and concave wingsails when the thrust coefficients are optimal, parallel simulations are conducted for the combination with the nose and flap at different cambers (as shown in Figure 21). Based on the results of a single evaluation of the nose and flap in the previous section, a range of 0 degrees to 20 degrees is determined to vary the interval of the nose camber and 40 degrees to 50 degrees to vary the interval of the flap camber, resulting in 25 scenarios for each wingsail.

In addition, since the variation trends of the thrust coefficient with a nose camber or flap camber are almost similar at different AOAs, a parallel study is carried out at an AOA of 8 degrees to simplify the analysis. The results indicate that the bare wingsail achieves a maximum thrust coefficient of 1.7 at a nose camber of 20 degrees and a flap camber of 50 degrees, while the concave wingsail reaches a maximum thrust coefficient of 2.1 at a nose camber of 15 degrees and a flap camber of 50 degrees, thus achieving an optimization of 23.5%. Furthermore, the comparison reveals that the thrust coefficients of the concave wingsail are higher than those of the bare wingsail in most of the folding configurations.

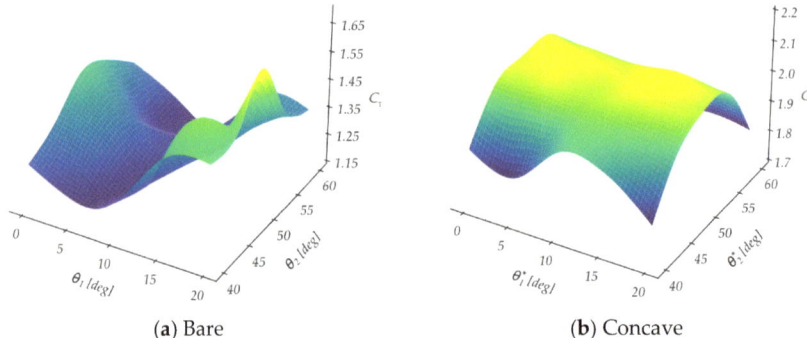

(a) Bare (b) Concave

Figure 21. Thrust coefficient of the wingsails at different cambers of both the nose and flap.

Through the above research, the optimal folding configuration of the wingsail is obtained. The camber angle of the bare wingsail's nose and flap is 20 degrees and 50 degrees, respectively, while for the concave wingsail, it is 15 degrees and 50 degrees, respectively. In order to provide a mechanical explanation of the performance difference, the pressure distribution around the wingsail is analyzed (as shown in Figure 22). The results show significant differences in pressure distributions on both the pressure and suction sides. On the pressure side of the bare wingsail, the airflow passes directly through the gap between the main body and flap, resulting in a reduction in the extent of the high-pressure area. Furthermore, the low-pressure area on the suction side of the concave wingsail is much more extensive and continuous, with distinct low-pressure areas occurring on both the suction side of the nose and main body. Consequently, the pressure difference between the two sides is greater, resulting in higher lift, particularly at an AWA of approximately 90 degrees, where the concave wingsail optimally utilizes wind energy, converting lift almost entirely into thrust. In particular, the suction side of the bare wingsail's flap also appears to have a significant low-pressure area. However, because this low-pressure area is located behind the flaps, it has little effect on lift, mainly leading to an increase in drag.

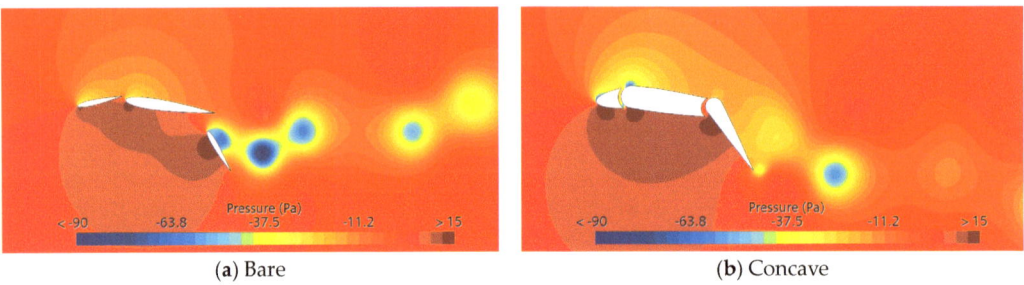

(a) Bare (b) Concave

Figure 22. Pressure distribution around the wingsails (AOA = 8 deg).

After obtaining the maximum thrust coefficient, the performance of the wingsail in utilizing wind energy to reduce the energy consumption of vessels can be further evaluated. As shown in Equations (5) and (6), V_v is the speed of the vessel, F_T is the thrust of the wingsail, P_w is the power of the wingsail, S is the surface area of the wingsail, and $C_{T\,max}$ is the maximum thrust coefficient. The specific values of the above parameters are shown in Table 3. For a wingsail with a chord length of 1 m and a unit span length, its surface area S is 1 m². Thus, the per unit wingsail surface area can provide an auxiliary propulsion power of 32.2 W per unit vessel speed. The calculations here consider the wingsails as an individual model and neglect the effects of waves in real sea states. In fact, the wingsails and deck are rigidly connected, which means the motion of the vessels will be affected by the coupling effect of wind and wave loads. This will disrupt the incoming airflow pattern, causing the wingsails to be unable to generate stable thrust and resulting in a certain loss of thrust power. In future studies, the coupling effect on the three-dimensional scale will be studied to reflect the auxiliary propulsion performance in a more comprehensive way.

$$P_w = F_T V_v \tag{5}$$

$$F_T = \frac{1}{2} C_{T\,max} \rho V_A^2 S \tag{6}$$

Table 3. Partial flow field parameters.

Parameter	Value
V_v	1 m/s
C_{Tmax}	2.1
ρ	1.225 kg/m³
V_A	5 m/s
S	1 m²

4. Conclusions

The performances of two types of three-element foldable wingsails, i.e., the original model (bare wingsail) and an optimized model (concave wingsail), are numerically investigated using the CFD method. By integrating quantitative performance metrics with an analysis of the physics governing wingsail aerodynamics, this work enables data-driven guidelines for balancing thrust capability, flow stability, and practicable engineering considerations in the design and application of wingsails, and the main conclusions are presented below.

- In an unfolded state, the aerodynamic and thrust performance of the concave wingsail is superior to that of the bare wingsail. In an AOA range of 4 to 10 degrees, the concave wingsail has a higher lift coefficient and lower drag coefficient, which results in a higher thrust performance for the same AOA and AWA. In addition, the flow pattern on the surface of the concave wingsail is consistently stable, with no significant vortex shedding, which indicates that the thrust performance is more stable.
- When evaluating the effect of the nose and flap cambers individually, it is found that rotating only the flap can significantly increase the thrust coefficients of both the bare and concave wingsails. However, it should be noted that the thrust coefficients decrease when the nose and flap cambers increase to certain critical values. In summary, the suitable variation interval for the nose and flap cambers are 0 to 20 degrees and 40 to 60 degrees, respectively.
- The thrust performance of both wingsails is further improved in the fully folded condition, i.e., when both the nose and flap are rotated. The maximum thrust coefficient of the bare wingsail is 1.7, when the nose's camber is equal to 20 degrees and the flap's camber is equal to 50 degrees. As for the concave wing, the maximum thrust coefficient is 2.1, at which the nose's camber is equal to 15 degrees and the flap's camber is

equal to 50 degrees. In particular, at an AOA of 8 degrees, the thrust coefficient of the concave wingsail is increased by 23.5% compared with the bare wingsail.

Author Contributions: Conceptualization, Y.J.; methodology, C.C.; software, C.C.; validation, T.C.; formal analysis, T.C.; investigation, C.C.; resources, Z.T.; data curation, Z.T.; writing—original draft preparation, C.C.; writing—review and editing, H.Y.; visualization, C.C. and H.Y.; supervision, Y.J. All authors have read and agreed to the published version of the manuscript.

Funding: This research received no external funding.

Data Availability Statement: Data is contained within the article.

Conflicts of Interest: The authors declare no conflict of interest.

Nomenclature

c	total chord of the wingsail [m]
c_1, c_1^*	chord of the nose of the wingsail [m]
c_2, c_2^*	chord of the main body of the wingsail [m]
c_3, c_3^*	chord of the flap of the wingsail [m]
g_1, g_1, g_1^*, g_2^*	gap of the wingsail [m]
Re	Reynolds number [-]
U	velocity of the inlet flow [m/s]
Δ_t	time step [s]
C_L	lift coefficient [-]
C_D	drag coefficient [-]
C_T	thrust coefficient [-]
V_A	apparent wind speed [m/s]
V_S	sailing wind speed [m/s]
V_T	true wind speed [m/s]
F_L	lift force [N]
F_D	drag force [N]
F_T	thrust force [N]
α	angle of attack [deg]
β	angle of apparent wind [deg]
θ_1, θ_1^*	camber angle of nose of the wingsail [deg]
θ_2, θ_2^*	camber angle of flap of the wingsail [deg]
x_1, x_1^*	pivot location of nose of the wingsail [-]
x_2, x_2^*	pivot location of nose of the wingsail [-]

References

1. Joung, T.H.; Kang, S.G.; Lee, J.K.; Ahn, J. The IMO initial strategy for reducing Greenhouse Gas (GHG) emissions, and its follow-up actions towards 2050. *J. Int. Marit. Saf. Environ. Aff. Ship.* **2020**, *4*, 1–7. [CrossRef]
2. Available online: https://www.dnv.com/news/imo-mepc-80-shipping-to-reach-net-zero-ghg-emissions-by-2050-245376 (accessed on 7 July 2023).
3. Nyanya, M.N.; Vu, H.B. Wind Propulsion Optimisation and Its Integration with Solar Power. Ph.D. Thesis, World Maritime University, Malmö, Sweden, 2019.
4. Julià Lluis, E. Concept Development of a Fossil Free Operated Cargo Ship. Ph.D Thesis, Chalmers University of Technology, Gothenburg, Sweden, 2019.
5. Available online: https://www.offshore-energy.biz/dnv-greenlights-oceanbirds-wing-sail/ (accessed on 29 August 2023).
6. Available online: https://www.prnewswire.com/news-releases/berge-bulk-unveils-the-worlds-most-powerful-sailing-cargo-ship-301955875.html (accessed on 16 October 2023).
7. Furukawa, H.; Blakeley, A.W.; Flay, R.G.; Richards, P.J. Performance of wing sail with multi element by two-dimensional wind tunnel investigations. *J. Fluid Sci. Technol.* **2015**, *10*, JFST0019. [CrossRef]
8. Schneider, A.; Arnone, A.; Savelli, M.; Ballico, A.; Scutellaro, P. On the use of CFD to assist with sail design. In Proceedings of the SNAME Chesapeake Sailing Yacht Symposium, Annapolis, MD, USA, 3 March 2003.
9. Li, C.; Wang, H.; Sun, P. Numerical investigation of a two-element wingsail for ship auxiliary propulsion. *J. Mar. Sci. Eng.* **2020**, *8*, 333. [CrossRef]

10. Li, C.; Wang, H.; Sun, P. Study on the Influence of Gradient Wind on the Aerodynamic Characteristics of a Two-Element Wingsail for Ship-Assisted Propulsion. *J. Mar. Sci. Eng.* **2023**, *11*, 134. [CrossRef]
11. Blakeley, A.W.; Flay, R.G.J.; Richards, P.J. Design and optimisation of multi-element wing sails for multihull yachts. In Proceedings of the 18th Australasian Fluid Mechanics Conference, Launceston, Australia, 3–7 December 2012.
12. Ma, R.; Wang, Z.; Wang, K.; Zhao, H.; Jiang, B.; Liu, Y.; Xing, H.; Huang, L. Evaluation Method for Energy Saving of Sail-Assisted Ship Based on Wind Resource Analysis of Typical Route. *J. Mar. Sci. Eng.* **2023**, *11*, 789. [CrossRef]
13. Shen, X.; Avital, E.; Rezaienia, M.A.; Paul, G.; Korakianitis, T. Computational methods for investigation of surface curvature effects on airfoil boundary layer behavior. *J. Algorithms Comput. Technol.* **2017**, *11*, 68–82. [CrossRef]
14. Bilandi, R.N.; Jamei, S.; Roshan, F.; Azizi, M. Numerical simulation of vertical water impact of asymmetric wedges by using a finite volume method combined with a volume-of-fluid technique. *Ocean. Eng.* **2018**, *160*, 119–131. [CrossRef]
15. Hassan, G.E.; Hassan, A.; Youssef, M.E. Numerical investigation of medium range re number aerodynamics characteristics for NACA0018 airfoil. *CFD Lett.* **2014**, *6*, 175–187.
16. Blount, H.; Portell Lasfuentes, J.M. CFD Investigation of Wind Powered Ships under Extreme Conditions. Master's Thesis, Chalmers University of Technology, Gothenburg, Sweden, 2021.
17. Sheldahl, R.E.; Klimas, P.C. *Aerodynamic Characteristics of Seven Symmetrical Airfoil Sections through 180-Degrees Angle of Attack for Use in Aerodynamic Analysis of Vertical Axis Wind Turbines (No. SAND-80-2114)*; Sandia National Lab (SNL-NM): Albuquerque, NM, USA, 1981.

Disclaimer/Publisher's Note: The statements, opinions and data contained in all publications are solely those of the individual author(s) and contributor(s) and not of MDPI and/or the editor(s). MDPI and/or the editor(s) disclaim responsibility for any injury to people or property resulting from any ideas, methods, instructions or products referred to in the content.

MDPI AG
Grosspeteranlage 5
4052 Basel
Switzerland
Tel.: +41 61 683 77 34
www.mdpi.com

Energies Editorial Office
E-mail: energies@mdpi.com
www.mdpi.com/journal/energies

Disclaimer/Publisher's Note: The statements, opinions and data contained in all publications are solely those of the individual author(s) and contributor(s) and not of MDPI and/or the editor(s). MDPI and/or the editor(s) disclaim responsibility for any injury to people or property resulting from any ideas, methods, instructions or products referred to in the content.

www.ingramcontent.com/pod-product-compliance
Lightning Source LLC
LaVergne TN
LVHW072346090526
838202LV00019B/2488